TERRA-1:
Understanding the
Terrestrial Environment

The Role of Earth Observations from Space

TERRA-1:
Understanding the
Terrestrial Environment

The Role of Earth Observations
from Space

Edited by
Paul M. Mather
Department of Geography
University of Nottingham, UK

Natural
Environment
Research
Council

BRITISH
AEROSPACE

CRC Press
Taylor & Francis Group
Boca Raton London New York

CRC Press is an imprint of the
Taylor & Francis Group, an **informa** business
A TAYLOR & FRANCIS BOOK

First published 1992 by Taylor & Francis Ltd

Published 2020 by CRC Press
Taylor & Francis Group
6000 Broken Sound Parkway NW, Suite 300
Boca Raton, FL 33487-2742

First issued in paperback 2020

ISBN 13: 978-0-367-57995-1 (pbk)
ISBN 13: 978-0-7484-0044-7 (hbk)

Visit the Taylor & Francis Web site at
http://www.taylorandfrancis.com

and the CRC Press Web site at
http://www.crcpress.com

British Library Cataloguing in Publication Data

A catalogue record for this book is available from the British Library.

Library of Congress Cataloging in Publication Data is available

Cover design by Amanda Barragry

Contents

Preface

Global environmental research is by its very nature world-wide in its scope and interdisciplinary in its applications. While oceanographers, meteorologists, climatologists and terrestrial scientists can, and do, carry out their individual fruitful lines of research, the interactions between the phenomena studied by these disciplines often fail to attract the attention they deserve. In recent years the concept of Earth system science has come to the fore; nowadays, emphasis is rightly placed on both the *analysis* of specific components of the Earth system and on the *synthesis* of the findings of individual disciplines in terms of how the Earth and its atmosphere function as an open system. A question of immediate importance is: how does this system change, both in the short (human) time-scale and over geological time, and how can such changes be predicted? Prediction, if it is to be reliable, requires scientific understanding which, in turn, requires models, theories and data. Remote sensing is capable of providing data on appropriate temporal and spatial scales. The problem is how best to encourage dialogue between terrestrial scientists and remote sensing experts.

An ad-hoc meeting organized on behalf of The Remote Sensing Society by Professor J.A. Allan and chaired by Lord Shackleton was held in August, 1989, at the Royal Geographical Society in London. The meeting provided the opportunity for a wide-ranging discussion on the role of remote sensing in continental and global scale monitoring and modelling of changes taking place on the Earth's land surfaces, and the effects of such changes on the global climate. Following this meeting the concept of an international conference on the theme of 'Understanding the terrestrial environment: the role of Earth observations from space' surfaced, and received wide support from a number of organizations and individuals. A steering committee, comprising Prof. R. Gurney (NUTIS, Reading University), Prof. J.-P. Muller (University College London), Dr B. Wyatt (Institute of Terrestrial Ecology), Prof. J.A. Allan (School of African and Oriental Studies, University of London) and Mr H. Mooney (British Aerospace (Space Systems) Ltd) was set up and, with a minimum of bureaucratic fuss, laid down the strategic aim of the conference: to bring together experts from those sciences which study the land surface of the Earth, including its form and features, and remote sensing specialists in order to promote the interchange of ideas and thus encourage inter- and multi-disciplinary interaction in facing the problems posed by environmental change.

As Chairman of the Steering Committee it was my duty to contact experts from many countries to invite them to TERRA-1, as speakers, session chairmen, or discussion leaders. I was pleasantly surprised by the keen interest expressed by all those whom I contacted, and had soon organized an impressive programme. Prof. P. Curran (University College, Swansea) provided me with considerable support at this time. The programme structure is reflected in the layout of this volume. The opening sessions of the conference were devoted to Earth and atmospheric science issues (including hydrological systems, geomorphology and soils, atmosphere–biosphere interactions, and biosphere). A session on sensor systems for global monitoring separated discussion of these scientific topics from the final part of the meeting, which was devoted to automated information extraction. The conference opened with addresses from leading scientists: Dr (now Sir) John Houghton, CBE, FRS, on 'Policy, implications of climate change'; Prof. W.G. Chaloner, FRS, on 'The IGBP—purpose and programmes'; and Mr M. Berens on 'NERC's TIGER programme'. Participants were also privileged to hear an after-dinner address on the social, environmental, political and economic ramifications of climate change from Sir Crispin Tickell.

A conference such as TERRA-1 does not occur spontaneously; moral and financial support is required. Two major sponsors came forward, British Aerospace PLC and the UK Natural Environment Research Council (NERC), and they deserve much of the credit for the success of TERRA-1. Dr B. Tinker, Director of Terrestrial and Life Sciences at the NERC, and Mr H. Mooney of British Aerospace gave invaluable assistance. I am particularly grateful to Mr Mooney and his team at Bristol for their help in every aspect of the organization of the meeting.

It is inevitable that I will have overlooked some of those who gave of their time and expertise, and I apologize where this is the case. I would particularly like to thank the Steering Committee, Prof. Curran, Dr Tinker, Mr Mooney, Mr D. Hardy and Ms K. Korzeniewski, the Remote Sensing Society's Administrative Secretary, also my wife Rosalind and daughter Tamsin for helping with the tedious task of sorting out registrations and accommodation. I am also grateful for the assistance of the session chairmen: Prof. B. Wilkinson (Institute of Hydrology), Prof. J.B. Thornes (University of Bristol), Prof. F. Becker (University of Strasbourg), Prof. P. Curran (University College, Swansea), Mr D. Hunt (British Aerospace), and Dr J. Dozier (University of California, Santa Barbara). The Chairman of the Natural Environment Research Council, Prof. J. Knill, and the Managing Director of British Aerospace (Space Systems) Ltd, Mr J. Holt, provided some very agreeable food for thought at the conference dinner. Finally, my thanks are due to speakers, chairman and delegates for attending the conference despite a difficult international situation preceding the war in Kuwait and Iraq. At one time it seemed likely that TERRA-1 would be a victim of international events, but in the end my unswerving belief in the toughness of terrestrial scientists carried me through to what, I hope, was the first of a biennial series of specialist conferences organized by The Remote Sensing Society on the general theme of global environmental change.

Paul M. Mather, Nottingham
November 1991

List of contributors

Prof. F. Becker/Dr Z. Li,
Ecole Nationale Superieure de Physique
de Strasbourg,
7 rue de l'Universite,
F-67000 Strasbourg,
France

Dr B. Choudhury,
NASA Goddard Spaceflight Center,
Greenbelt,
Maryland MD 20771,
USA

Dr C.S.M. Doake,
British Antarctic Survey,
High Cross,
Madingley Road,
Cambridge CB3 0ET

Dr J. Dozier,
NASA Goddard Spaceflight Center,
Code 900,
Greenbelt,
Maryland MD 20771,
USA

Dr P. Francis,
Dept. of Earth Sciences,
The Open University,
Walton Hall,
Milton Keynes,
MK7 6AD

Prof. R. Gurney,
NERC Unit for Thematic Information
Systems,
Dept. of Geography,
The University,
Whiteknights,
P.O. Box 227,
Reading RG6 2AB

Dr B.L. Isacks,
INSTOC,
Dept. of Geological Sciences,
2122 Snee Hall,
Cornell University,
Ithaca, NY 14853
USA

Prof. P.G. Jarvis/Dr J. Moncrieff,
Dept. of Forestry and Natural
Resources,
Darwin Building,
The King's Buildings,
Mayfield Road,
Edinburgh EH9 3JU

Dr J.-P. Malingreau,
Commission of the European
Communities,
Joint Research Centre,
Building 44,
I-21020 Ispra (Varese),
Italy

Prof. P.M. Mather/Dr M.D. Steven,
Geography Department,
University of Nottingham,
Nottingham NG7 2RD

Dr Bernard Pinty,
LERTS,
18 ave. E. Belin,
F-31055 Toulouse,
France

Dr S. Quegan,
Dept. of Applied Mathematics,
The University,
Sheffield S10 2TN

Prof. C. Rapley,
Mullard Space Science Laboratory,
Dept. of Physics and Astronomy,
University College London,
Holmbury St Mary,
Dorking,
Surrey RH5 6NT

Dr M. Rast,
European Space Agency,
ESTEC,
Noordwijk,
The Netherlands

Dr P.R. Rowntree,
The Hadley Centre,
Meteorological Office,
London Road,
Bracknell RG12 2SZ

Dr Anatolij M. Shutko,
Academy of Sciences of the USSR,
Institute of Radio Engineering and
Optics,
K. Marx Av. 18,
Moscow GSP 3,
103907 USSR

Dr W.J. Shuttleworth,
Institute of Hydrology,
Maclean Building,
Crowmarsh Gifford,
Wallingford,
Oxon OX10 8BB

Dr B.K. Wyatt,
Environmental Information Centre,
The Institute of Terrestrial Ecology,
Monks Wood Experimental Station,
Abbots Ripton,
Huntingdon PE17 2LS

List of acronyms

AATSR	Advanced Along Track Scanning Radiometer
AMI	Active Microwave Instruments
AMSU (MTS/MHS)	Advanced Microwave Sounding Unit (Microwave Temperature Sounder/Microwave Humidity Sounder)
ARGOS	Data Collection and Location System
ATSR	Along-Track Scanning Radiometer
ARME	Amazon Region Micrometeorological Experiment
AVHRR	Advanced Very High Resolution Radiometer
BAHC	Biological Aspects of the Hydrological Cycle
BOREAS	Boreal Ecosystem Atmosphere Study
BRDF	Bidirectional Reflectance Distribution Function
CERES	Clouds and the Earth's Radiant Energy System
CORINE	Co-ORdinated INformation on the Environment in the European Community
DMSP	Defense Meteorological Satellite Program
EOS	Earth Observing System
EOSDIS	Earth Observing System Data and Information System
ERS	Earth Resources Satellite
ESA	European Space Agency
FIFE	First ISLSCP Field Experiment
GAC	Global Area Coverage
GEMS	Global Environment Monitoring System
GIS	Geographical Information System
GCIP	GEWEX Continental-scale International Project
GCM	General Circulation Model
GEWEX	Global Energy and Water Cycle EXperiment
GOMOS	Global Ozone Monitoring by Occultation of Stars
GRID	Global Resource Information Database
HAPEX	Hydrological & Atmospheric Pilot EXperiments
HAPEX-MOBILHY	Hydrological & Atmospheric Pilot EXperiments/ MOdelisation du BILan HYdrique
IASI	Infrared Atmospheric Sounding Interferometer
IGBP	International Geosphere-Biosphere Programme
IRTS	Infra-Red Temperature Sounder
ISLSCP	International Satellite Land Surface Climatology Project
JERS	Japanese Earth Resources Satellite
LOTREX	LOngitudinal Land-Surface Transverse EXperiment

MCP	Meteorological Communications Package
MERIS	MEdium Resolution Imaging Spectrometer
MIPAS	Michelson Interferometer for Passive Atmospheric Sounding
NASA	National Aeronautics and Space Administration (USA)
NERC	Natural Environment Research Council (UK)
NOAA	National Oceanographic and Atmospheric Administration (USA)
POEM	Polar Orbit Earth Observation Mission
PRARE	Precise Range And Range Rate Experiment
PRAREE	Precise Range And Range Rate Experiment Extended version
RA-2	Radar Altimeter-2
S&R	Search and Rescue Package
SAR	Synthetic Aperture Radar
SCATT	Wind SCATTerometer
SCIAMACHY	Scanning Imaging Absorption Spectrometer for Atmospheric Cartography
SMMR	Scanning Multichannel Microwave Radiometer
SSM/I	Special Sensor Microwave Imager
SVATS	Soil/Vegetation/Atmosphere Transfer Schemes
TIGER	Terrestrial Initiative into Global Environmental Research
TOMS	Total Ozone Mapping Spectrometer
VIRSR	Visible and InfraRed Scanning Radiometer
WRCP	World Climate Research Programme

Chapter 1
Remote sensing the land surface water budget

R.J. GURNEY

NERC Unit for Thematic Information Systems,
University of Reading

Introduction

The study of the land surface as part of the climate system is important because of the role of the land surface in radiation interception, in partitioning energy and water fluxes, and because it is the part of the climate system with the best and longest data records. The land surface is also important for economic reasons through impact of possible changes in climate.

The energy budget at the land surface may be written:

$$R_N - G = L.E + H$$

where R_N is the net radiation, G is the soil heat flux, $L.E$ is the latent heat flux, and H is the sensible heat flux. Similarly, the water budget at the land surface may be written:

$$P - E = Q + \Delta S$$

where P is the precipitation, E is evaporation, Q is discharge and ΔS is change in storage in soil or groundwater. Evaporation is linked to the energy budget because of the latent heat required, and may be evaporation from vegetation, from the soil directly, or from water standing on leaves after rain, commonly called the interception loss.

These basic equations have been well known for perhaps two centuries, and observations and models have been made to study the energy and water budgets in small areas. However, these observations are patchy and there are large parts of the world where the uncertainties in the energy and water budgets at the land surface are very large and of undetermined size. The very large spatial variability of the land surface, in topography, soil and vegetation properties and geology, means that methods of averaging sparse observations over large areas are very unreliable. General circulation models, which may be used to study the interactions of the land surface with the rest of the climate system, only contain the most rudimentary land surface representations that cannot easily be checked against observations, again largely because of scale considerations (Sellers *et al.*, 1986).

Remotely-sensed data provide good spatial and temporal coverage of the
land surface and can be used for retrieving variables related to components of
the energy and water budgets. These variables can be used in models to
interpolate the energy and water budgets between conventional observations. The methods involve confronting the models with several types of
data, possibly including several types of remotely-sensed data. The model
parameters also have to be averaged spatially, and this is not necessarily
straightforward where non-linear processes are involved. The difficulties
involved mean that progress has been slower than in atmospheric or ocean
remote sensing. Much effort in land remote sensing has been devoted to
more empirical solutions than those required here, such as mapping of land
cover types or areas of tropical deforestation. We need radiative transfer
models which relate the radiation observed to the radiation loading at some
part of the surface, coupled to geophysical models of the surface energy and
water budgets which are closely linked and must be solved simultaneously.
There are usually more unknown parameters than observations and the
model equations are non-linear, so some maximum likelihood method of
solution has to be found. The processes are fairly conservative though, so the
problem is not as ill-determined as it may appear.

The next section summarizes some of the recent work in this area. Much
of this work has been carried out as part of large interdisciplinary field
experiments attempting to estimate areal budgets. The following section
considers how the observations are put together in models over areas.

Surface energy budget estimation

The radiation budget is a critical part of the energy budget. In principle,
remote sensing can be of great value because it is only radiation that is
measured at a sensor. The surface radiation budget consists of three terms:

$$R_N = (1 - a_s)R_s\!\downarrow + R_L\!\downarrow - R_L\!\uparrow$$

where a_s is the albedo, $R_s\!\downarrow$ and $R_L\!\downarrow$ are the short- and long-wave incoming
radiation respectively, and $R_L\!\uparrow$, the long-wave outgoing radiation. A satellite
sensor estimates the outgoing radiation at the edge of the atmosphere, so at
visible and infrared wavelengths corrections must be made for atmospheric
transmission. No current sensor makes sufficient simultaneous observations of
the atmosphere and surface to allow such observations to be made.

The reflectance of the land surface varies with angle in a complex way,
depending on surface geometry and solar azimuth angle. The albedo of the
land surface is determined by the reflectance of the surface components and
their geometry, and therefore varies with angle of illumination. It can also
not be well estimated, therefore, from a single view of a scene by a remote
sensor. The Bidirectional Reflectance Distribution Function (BRDF) can,

however, be well estimated by only a few off-nadir observations if they are combined with a nadir view, even with one off-nadir and one nadir view if the off-nadir view is in the solar principal plane (Camillo, 1987). The BRDF is necessary for estimating albedo, and so almost simultaneous nadir and off-nadir observations are required if it is to be estimated.

Atmospheric effects change rapidly in time. In particular, clouds have an effect on the surface radiation budget. It is therefore necessary to have observations of cloud extent, cloud thickness and composition and cloud base height, with as much of this information as possible being obtained from geostationary platforms. However, the sensors which estimate the surface BRDF and other surface features must make simultaneous observations of atmospheric transmission in order to correct the remotely sensed data for atmospheric effects. It should be noted that no joint retrievals of atmospheric transmission and surface bidirectional properties have yet been routinely successful. This is an area of continuing research.

The long-wave budget has great uncertainties, because of the magnitude of the incoming radiation and of atmospheric effects, and because of the variability of surface emissivity, particularly with varying water content, that affects the infrared brightness temperature. There must therefore be a joint retrieval of atmospheric emission, surface emission and surface emissivity. The only way this can be achieved is by using a high spectral resolution infrared radiometer. This type of instrument, however, has severe problems of spatial resolution. Nevertheless, such an instrument is necessary for progress to be made.

At both short and long wavelengths, the surface radiation budget varies rapidly in time, with solar illumination and cloud cover. For climate studies, it is important that those parameters related to the radiation budget be observed; the instantaneous radiation balance itself can only be observed occasionally when a satellite is overhead. The choice of instruments used to observe the BRDF and emissivity of the surface is therefore particularly important, in order to obtain the prescribed variables in surface radiation budget models. The instantaneous radiation budget, including the effects of clouds and other atmospheric effects, and illumination, can then be modelled in a general circulation model (GCM), with occasional checking against actual observations when these are available. This procedure will be challenging to establish and is not included in existing surface models in GCMs. Although the existing surface models include radiation components, they are rarely estimated from observation.

Energy and water budget

The energy and water budgets have to be considered together, as explained above. Several state variables related to these budgets can be observed with remote sensing.

As explained earlier, surface temperature can be observed with remote sensors, although considerable increases in accuracy are required. Several workers have suggested calibrating energy budget models using remotely sensed surface temperature, although experimental work has shown that the aerodynamic and radiative temperatures are different for vegetated surfaces (for example, Huband and Monteith, 1986). However, the differences between surface and atmospheric temperatures are small, so the possible errors in estimating sensible and latent heat fluxes based on these differences are prone to large errors, even to having the wrong sign.

Temperature and moisture profiles in the atmosphere may be assimilated into numerical weather prediction models. The large-scale divergence of water vapour or sensible heat near the surface can be used to estimate the large-scale surface evaporation or sensible heat fluxes. These large-scale estimates can in principle be apportioned to local areas given good local atmospheric and land surface models. For such a soil–vegetation–atmosphere transfer model, we need an estimate of the store of available moisture. Surface soil moisture can be closely and replicably estimated for surfaces with agricultural crops or grass, using passive microwave 1.4 GHz radiometry (Schmugge, 1983). The potential uses of such data are numerous and extend beyond the topic of this paper, however, there are limitations—the spatial resolution from spaceborne platforms and the methods of relating the spatial distribution of surface soil moisture, in the top 5 cm of the soil column, to the moisture in a whole soil profile. As with other data sets, these limitations will probably be overcome by assimilating the data into soil–vegetation–atmosphere transfer models in GCMs.

Camillo *et al.* (1986) used surface temperature and L-band passive microwave time series to retrieve hydraulic properties for a variety of soils. Almost as a by-product, time series of evaporation and soil moisture profiles were produced. They wrote a coupled energy and water budget model for the upper soil, with a diffusion model of the atmosphere near the surface. Given initial estimates of soil and atmospheric properties and a time series of atmospheric observations, time series of moisture profiles could be produced for the three soils studied in the experimental site. The moisture profiles were then used in a radiative transfer model to estimate microwave brightness temperatures that could then be compared with observations from a truck-mounted microwave radiometer. The hydraulic parameters could then be adjusted, and the procedure repeated, until the observed and estimated remotely-sensed observations agreed. This approach is similar to that adopted in assimilation of data into numerical weather forecast models. A similar approach will probably be adopted to estimate profile moisture routinely once suitable passive microwave observations are available.

Snow area and volume are hydrological variables which have been observed from space for many years and for which hemispheric observation sets are now routinely available (e.g. Chang *et al.*, 1987). Snow area can be routinely observed with visible sensors because of the very different albedo of snow areas compared with that of snow-free areas. Snow volume may be estimated by using the microwave emission from the snowpack, and comparing this with the emission from the underlying soil measured at a longer wavelength. A simple radiative transfer model can be used, which assumes a defined average snow grain size, usually 0.3 mm. Both Special Sensor Microwave Imager (SSM/I) and Scanning Multichannel Microwave Radiometer (SMMR) data have been used to derive monthly snow volume data since 1979; for SMMR, 18 and 37 GHz horizontal polarization data are used, with similar wavelengths for SSM/I (Foster *et al.*, 1987).

Vegetation area and perhaps vegetation status have been observed using a variety of data, including vegetation indices and microwave polarization differences. The ways these data can be used in energy budget models are still very unclear, and work is needed in the improvement of vegetation representation in these models, particularly at global scales. Although there has been considerable work on vegetation representation in energy budget models, and on mapping vegetation from space, there has been little study of the two areas of work together.

Data integration and spatial scale

Data requirements for 1-D energy budget models were discussed in very general terms in the preceding section. It is very unclear how model parameters are related in point and areal average energy budget models. Several experiments have been conducted to address the problems of spatial scale and how remote sensing can be used to calibrate spatially-averaged energy budget models. The best known are the HAPEX-MOBILHY† (André *et al.*, 1986) and FIFE† (Sellers *et al.*, 1988) studies, in south-western France and Kansas, respectively. Future experiments include HAPEX-Sahel, in Niamey, and BOREAS†, in Canada, for semi-arid grassland and boreal forest respectively. Results from these experiments generally show that simple spatial averaging can be used in energy budget models, but much further work is needed on how remotely sensed data may be used.

Shuttleworth *et al.* (1989) have shown that, on the FIFE site, the evaporative fraction, the ratio of the latent heat flux to the available energy, the sum of the latent and sensible heat fluxes, is very constant during the day on cloudless days, for a range of surface flux sites. The observations were from several types of eddy correlation and Bowen ratio device. There were consistent differences during a given day between sites, but the differences varied from day to day, and bore little relation to the vegetation or topography at the site. Further studies indicated that while there were spatial variations in the actual evaporation at a given time, these were sufficiently

---

†HAPEX–MOBILHY Hydrological and Atmospheric Pilot Experiments/Modelisation du Bilan Hydrique; FIFE First ISLSCP Field Experiment; BOREAS Boreal Ecosystem Atmosphere Study.

random for a simple spatial average of the observations to be taken, at least for this relatively simple 15 × 15 km prairie grassland. It will be very interesting to compare these results with those for the much more moisture-limited and spatially heterogeneous savannah grassland of the HAPEX-Sahel experiment. Again, early modelling work (e.g. Shuttleworth and Gurney, 1990) has shown that fairly simple extensions to existing models may be used to apply remotely-sensed data for areas of partial vegetation cover.

Wang et al. (1989) provide another example where the surface heterogeneity appears to be very large but where simple modelling averages the variability satisfactorily. A time series of the spatial variability of surface soil moisture was observed using an L-band airborne radiometer on an instrumented catchment at the FIFE site after a rainstorm. The spatial heterogeneity of the time series is very great and apparently bears little relationship to topography. However, Wang et al. (1989) were able to fit a very simple hill slope model and calibrate it with average soil moisture, as observed, thus closely reproducing the observed streamflow. This is a further indication that simple averaging may be appropriate.

Conclusions

Several conclusions may be drawn from this brief summary of radiation and energy budget studies for climate purposes:

(1) Simultaneous nadir and off-nadir observations are needed at visible and near-infrared wavelengths to characterize Bidirectional Reflectance Distribution Functions, with simultaneous observations for atmospheric correction.
(2) A higher spectral resolution thermal infrared radiometer is needed to make atmosphere moisture and temperature observations, and also to make estimates of surface emittance and temperature.
(3) Passive microwave observations at 1.4 GHz are needed to estimate near-surface soil moisture.
(4) Passive microwave observations at approximately 18 and 37 GHz are needed to estimate snow volume.
(5) Simple visible and near infrared spectral observations at several different wavelengths are needed to map vegetation. Simultaneous atmospheric correction is also required.
(6) Experiments to study scaling-up of energy budget relationships to large spatial scale are required for algorithm development and to define satellite radiometer requirements.
(7) The methods by which all the remotely-sensed observations are jointly used to retrieve surface energy and water budgets are still very preliminary and need further work.

References

André, J.-C., Goutourne, J.-P. and Perrier, A., 1986, HAPEX-MOBILHY: A hydrologic-atmospheric experiment for the study of water budget and evaporation flux at the climatic scale, *Bulletin of the American Meteorological Society*, **67**, 138–44.

Camillo, P.J., 1987, A canopy reflectance model based on an analytic solution to the multiple scattering equation, *Remote Sensing of Environment*, **23**, 453–77.

Camillo, P.J., O'Neill, P.E. and Gurney, R.J., 1986, Estimating soil hydraulic parameters using passive microwave data, *IEEE Transactions Geoscience and Remote Sensing*, **GE-24**, 930–6.

Chang, A.T.C., Foster, J.L. and Hall, D.K., 1987, NIMBUS-7 derived global snow cover parameters, *Annals of Glaciology*, **9**, 39–44.

Foster, J.L., Hall, D.K. and Chang, A.T.C., 1987, Remote sensing of snow, *EOS, Transactions of the American Geophysical Union*, **68** (32), 681–4.

Huband, N.D.S. and Monteith, J.L., 1986, Radiative surface temperature and energy balance of a wheat canopy. *Boundary Layer Meteorology*, **36**, 1–17.

Schmugge, T.J., 1983, Remote sensing of soil moisture: Recent Advances, *IEEE Transactions Geoscience and Remote Sensing*, **GE-21**, 336–44.

Sellers, P.J., Mintz, Y., Sud, Y.C. and Dalcher, A., 1986, A simple biosphere model for use within general circulation models, *Journal of Atmospheric Sciences*, **43**, 505–31.

Sellers, P.J., Hall, F.G., Asrar, G., Strebel, D.E. and Murphy, R.E., 1988, The first ISLSCP field experiment, *Bulletin of the American Meteorological Society*, **69**, 22–7.

Shuttleworth, W.J. and Gurney, R.J., 1990, The theoretical relationship between foliage temperature and canopy resistance in sparse crops, *Quarterly Journal of the Royal Meteorological Society*, **116**, 497–519.

Shuttleworth, W.J., Gurney, R.J., Hsu, A.Y. and Ormsby, J.P., 1989, FIFE: The variation in energy partition at surface flux sites, in Rango, A. (Ed.) *Remote sensing and large-scale global processes*. Wallingford, England: IAHS Publication 186, pp. 67–74.

Wang, J.R., Shiue, J.C., Schmugge, T.J. and Engman, E.T., 1989, Mapping soil moisture with L-band radiometric measurements, *Remote Sensing of Environment*, **27**, 305–12.

Chapter 2

The use of visible satellite imagery over ice sheets

D.G. VAUGHAN and C.S.M. DOAKE

British Antarctic Survey,
Cambridge

Introduction

The ice sheets of Antarctica and Greenland modulate world climate and control global sea level. Changes in snow and ice extent alter the Earth's albedo and consequently the radiation budgets which drive the atmospheric engine. Ice sheets influence the transfer of energy, mass and momentum between land, ocean and atmosphere and control weather patterns on many temporal and spatial scales. Because the polar ice sheets are inaccessible and inhospitable areas of the Earth's surface, both their historical and contemporary behaviour are poorly mapped and understood. For many parts of the globe, satellite data offer not only the chance to observe large areas synoptically, but also an alternative perspective to ground based observations. Over the hinterlands of Greenland and Antarctica, however, satellite derived data are for the most part, all that are available. In 1977, when the use of satellite imagery was still rare, Swithinbank (1977) wrote that the far side of the Moon was much better mapped than parts of our own planet and noted '... that position errors of more than 100 km were found in 1975 on the most up-to-date maps published at any scale of one part of the Antarctic Peninsula'.

Ice sheets are dynamic systems; snow falling on the ice sheet is compressed and advected towards the margins, at speeds in excess of 1000 m per year in some areas. Ice is lost by the processes of iceberg calving and surface and basal melting, establishing a possibly precarious balance between accumulation and wastage. The flux passing into and out of the Antarctic Ice Sheet system is estimated to be around 2000 Gt per year (Warrick and Oerlemans, 1990), corresponding to around 5 mm per year of global sea level.

A recent synthesis of studies of global sea level rise suggests a rate of 1-2 mm per year since the beginning of the century (Warrick and Oerlemans, 1990). Although most of this rate is accounted for by thermal expansion of the ocean and wastage of mountain glaciers, melting of the polar ice sheets could give rise to a considerably more rapid and eventually larger sea-level rise; the Antarctic Ice Sheet holds 90 per cent of the world's ice, the Greenland Ice Sheet holds around 9 per cent, and only 1 per cent is held in the smaller ice caps and mountain glaciers. Most general circulation models of world climate agree that the equilibrium temperature response to increasing CO_2 is likely to be pronounced in polar regions, and so even a modest rise in mean global temperature could cause a significant change in polar ice sheets.

The fundamental question most asked of glaciologists is: are the polar ice sheets growing or shrinking and how will they change in the future? As our understanding of the dynamics of ice sheets improves, glaciologists are becoming increasingly aware that to answer this question ice flow must be studied as a complex system of external forcings, internal triggers and feedback mechanisms and local studies cannot be extrapolated easily over the whole ice sheet. For example, the flow of some ice streams appears to be controlled primarily by glacio-geologic processes at the glacier bed and to be subject to substantial change on 100 year timescales (Shabtaie and Bentley, 1987). Beyond the question of sea level change, predictions of the integrated mass balance and total extent of the great ice sheets will become vital as models of global climate and oceanic circulation demand more precise boundary conditions, and more reliance is placed in these models by policy makers.

Discussion of sensors

Visible imagery has proved to be among the best tools at the disposal of glaciologists studying the dynamics of ice sheets, since it provides a detailed picture of a variety of features produced by and controlling ice flow.

High-resolution imagery

On 23 July 1972 NASA launched the first Earth Resources Technology Satellite (ERTS-1), its multi-spectral scanners (MSS) giving glimpses of hitherto unexplored regions of the world with 80 m resolution. Glaciologists were not slow in realizing the great potential of the instrument to save them time and expense in reaching isolated areas (Krimmel and Meier, 1975; Østrem, 1975). ERTS-1, later renamed LANDSAT 1, was followed by LANDSATs 2–5. LANDSATs 4 and 5 carried, in addition to the MSS, the thematic mapper, a 7 band visible and infrared sensor with a nominal ground resolution of 30 m.

The French commercial satellite SPOT (Système Probatoire d'Observation de la Terre) launched in 1986, has improved on LANDSAT resolution. In a panchromatic mode it can achieve 10 m resolution. Features as small as individual crevasses can now be resolved, but with a consequent reduction of the image width to only 60 km compared to 180 km for LANDSAT.

The LANDSAT series can obtain imagery at latitudes up to 82.6°, whereas SPOT with its off-nadir viewing capability can obtain images at latitudes up to 86°. Off-nadir viewing requires pre-programming of the satellite which, together with the likelihood of the images being unusable for glaciological studies because of cloud cover, often makes them prohibitively expensive.

Low-resolution imagery

Low-resolution imagery, available from meteorological satellites, covers the whole of Antarctica several times each day, giving a much greater chance of obtaining cloud-free images. The potential for using this type of imagery was realized by Bindschadler and Vornberger (1990), and by Casassa *et al.* (1991). They used imagery from the advanced high-resolution radiometer (AVHRR), different models of which have been included on NOAA satellites since 1978. This sensor has one channel in the visible spectrum and either three or four in the short-wave and thermal infrared. AVHRR has a swath of about 2500 km and a resolution of 1100 m at nadir. Bindschadler and Vornberger showed that AVHRR captures many flow features previously seen on higher resolution images. Areas of grounded, floating, stagnant and streaming ice sheet could be distinguished, together with streamlines and ice divides. Glaciologists have perhaps been slow to exploit low-resolution imagery, from AVHRR and the Defense Meteorological Satellite Program (DMSP) satellites.

Photographic imagery

High resolution (6 to 20 m) photographic images taken by the Soviet Soyuz satellites since 1976 are now available to Western scientists. Although a photographic image cannot be analysed as easily as digital data, this is a promising source of data that should not be overlooked (Ferrigno and Molnia, 1989).

Technical considerations

Snow fields provide the most homogeneous reflecting surfaces on Earth, that is to say they are continuously white over very large areas. This homogeneity means that very subtle changes in surface topography are largely responsible for prominent patterns on images. Although topographic signals are present in images of other areas of the world, they are normally obscured by differences in surface albedo and colour due to differing surface composition.

Saturation

Few satellite sensors were designed specifically to cope with the problems of viewing snow and ice. For visible imagery a major problem arises as the albedo of clean snow is around 0.8 or 0.9, much higher than the 0.1 to 0.3 more typical of terrestrial surfaces. Thus when the sun elevation is high, snow often reflects enough energy to saturate the sensors (Dowdeswell and McIntyre, 1986). This problem can be solved by obtaining imagery taken with a sun elevation of less than 20°, but for satellites in sun-synchronous orbits this means that usable images can only be obtained at high latitudes during the spring and autumn.

Image enhancement

The homogeneity of snow cover reflectance leads to images having very low contrast. On an ice sheet even good LANDSAT images may often span fewer than 10 brightness values. The visibility of features may be improved by contrast stretching but with an attendant increase in noise levels. The spectral response of snow is however, almost uniform across the visible and near-infrared bands, i.e. 0.3-1.0 μm (Choudry and Chang, 1981); in other words, it is white! Each band contains very similar information. A number of workers (e.g. Orheim and Luchitta, 1987) have used this property to enhance the topographic information by constructing the composite image via principal component analysis. The first principal component (PC1) is assumed to give mostly topographic information since it contains the signal common to all the images. Other components contain information on inter-band reflectance differences.

An alternative method of concentrating topographic information into a single image is to combine the visible bands in the ratio specified by the absolute calibrations of their sensors (Vaughan et al., 1988). The resulting image is thus related to the total energy incident on the sensor. Incoherent noise within the sensors is diluted by this method and quantization effects reduced. This process also has the advantage of consuming considerably less computing time.

Cloud recognition

The similarity of reflectivities for snow and clouds also causes problems of differentiation. Automatic schemes to do this classification, based on inter-channel differences or spatial coherency tests have met with limited success (Turner, 1987; Gessel, 1989). The most difficult cloud types to identify are uniform layers of cirrus with cloud top temperatures close to those of the snow surface. This presents a major difficulty for users, since the institutions disseminating imagery usually do not properly distinguish cloud from ice, and when images classified as cloud-free are ordered, they rarely turn out to meet specification. Conversely, many usable images are hidden away after being erroneously classified as having significant cloud cover.

Uses of imagery

Planimetric mapping

Pre-satellite maps of Antarctica often consisted of large areas of blank paper sparsely intersected by the tracks of sledge-borne and aerial traverses. In 1978 the British Antarctic Survey published a series of maps of the Antarctic Peninsula at 1:250 000 produced from ERTS-1 data. These maps were little more than annotated reproductions of the original photographic product overlaid with an approximate graticule, but even today they remain the best available in some areas. Nowadays field workers often rely on maps derived almost entirely from visible imagery (e.g. Swithinbank *et al.*, 1988; Grosfeld *et al.*, 1989). Expensive and time-consuming aerial photography is now rarely justified except for the largest scale mapping.

Ice velocity mapping

Crevasses are surface markers that are advected at the same speed as the ice, allowing ice velocity maps to be constructed from repeat images (e.g. Luchitta and Ferguson, 1986; Orheim and Luchitta, 1987). Vaughan *et al.* (1988) derived similar measurements from moving thickness variations in Ronne Ice Shelf. Velocity maps are usually produced from repeat imagery by rectification to fixed points within the image, but recently an alternative method has been developed by Bindschadler and Scambos (1991). Two images of ice sheet with no rock outcrops were overlaid using as reference the long-wavelength ice surface topography which reflects the bedrock topography and so is stationary in relation to it. Crevasses moving with the ice were used as markers with which to determine ice velocity. This method has great potential and is likely to be widely used.

Iceberg tracking

Satellite images have been used since 1969 (Swithinbank *et al.*, 1977) to chart the drift of icebergs. A recent study by Keys *et al.* (1990) of one of the largest icebergs on record (154 × 35 km) used LANDSAT, DMSP, and NOAA imagery to interpret the calving event and describe the subsequent drift. Tracking of large icebergs provides useful information on ocean currents, thickness of ice and iceberg interactions with sea ice and the sea bed.

Topographic mapping: photoclinometry

Surface topography of the planets has been calculated using the radarclinometric method (Wildey, 1984) which uses the radiometric variation in a single image and some formulation of the bidirectional reflectance function to calculate the surface slopes. Vaughan *et al.* (1988) showed that this method could be applied to the homogeneous snow surfaces of Rutford Ice Stream, Antarctica, and more recently Casassa (personal communication, 1990) had considerable success in reproducing imagery from topographic contours on Ice Stream C using a Lambertian reflectance function. Unfortunately, the inverse problem (imagery to topography) is not well constrained and requires for solution more than a single image and an estimate of the bidirectional reflectance function. Two images of the same area could be processed together if they had significantly different sun azimuths, not easy to obtain at high latitudes. Alternatively, ground truth within the region might be used to control the solution. Significant work remains to be done on the method, but it is potentially a useful way of producing small-scale terrain models.

Glacio-geologic mapping of former ice sheets

Since a major part of our interest in ice sheets concerns their influence on and response to contemporary climate change, it is pertinent to ask how the ice sheets have altered during historical climate changes. The last Ice Age ended a mere 10 000 years ago, with the retreat of the Laurentide Ice Sheet which had covered much of the North American Continent during the last glacial period. The collapse of this ice sheet was remarkably quick and coincided with substantial alteration of climate, although whether as cause or effect is still uncertain. Boulton and Clark (1990) have used LANDSAT imagery to map drift lineations in North America. From this they produced a 'tentative time-space envelope' for the Laurentide Ice Sheet showing its approximate extent over the last glacial cycle, 120 000 years.

Interpretation of ice dynamics

High-resolution imagery shows many glaciological features important to an understanding of the dynamics of ice sheets.

(1) *Flow lines* (Crabtree and Doake, 1980), alternatively named *fossil flow lines* (Swithinbank and Luchitta, 1986), and *plumes* or *streak lines* (MacAyeal *et al.*, 1988), are some of the most useful features available to glaciologists in visible imagery. These linear features appear to record the path along which the ice has moved. They appear on ice sheets but are most noticeable on ice shelves, often becoming visible at the grounding line and persisting right up the ice front, a lifespan of many centuries. It is likely that flow lines on images are the expression of topographic ridges and troughs, but the amplitude of the variations may be very small. Optical levelling over a flow line, clearly visible on MSS imagery, showed the topography to be of the order of 1 m in 1 km (0.1 per cent slope) (A. Smith, personal communication). The longevity of flow lines is thus perhaps surprising; since annual snow accumu-

lation can be around 1 m, one might expect such small features to become buried quickly. The implication is, perhaps, that accumulation is highly homogeneous over the ice shelf, or that there is a material difference in the properties of the ice associated with the flow line.

(2) *Crevasses* are fractures in the surface of an ice mass caused by excessive stress associated with flow. On cold ice sheets they can be up to 100 m wide and may be bridged for much of their lives by falling and drifting snow. Although only the best high-resolution imagery can distinguish individual crevasses, areas of crevassing can often be observed as a change in image texture. Crevasses are particularly important since they are a quantifiable indication of the direction and magnitude of stresses in the ice mass (D.G. Vaughan, submitted), giving a direct comparison with the results of numerical models of ice sheet dynamics.

(3) *Grounding lines* separate grounded ice sheets from floating ice shelves. Since the melting of floating ice does not contribute directly to sea-level, any change in sea-level due to change in ice sheet size will almost certainly be accompanied by significant grounding line displacement. Grounding line movement is thus of primary concern to those making estimates of future mass balance. Even where the ice is over 1000 m thick, grounding lines can usually be located within a kilometre by a characteristic change in texture on visible imagery. A particularly striking example was given by Swithinbank *et al.* (1988) for Recovery Glacier, Antarctica. As ice flows over bedrock the surface develops undulations that are the reflection of those in the bedrock. These undulations are very quickly lost when they become unsupported on the ice shelf.

(4) *Ice streams* are large fast flowing glaciers that drain approximately 90 per cent of the ice from the Antarctic Ice Sheet and supply most of the ice to the ice shelves. As the most dynamic element of the ice system, they are the focus of much attention and fieldwork. As an example consider Rutford Ice Stream and Carlson Inlet, Antarctica which have similar surface gradients and thicknesses, and hence similar driving stresses. LANDSAT imagery has, however, shown Rutford Ice Stream to have the surface texture typical of streaming flow, whereas Carlson Inlet appears very smooth and hence quiescent. These inferences were confirmed by surface measurements which showed the velocities to be over 300 m per year for Rutford Ice Stream and less than 10 m per year for Carlson Inlet (Frolich *et al.*, 1989).

(5) *Ice rumples* are areas where an ice shelf becomes grounded on a shoal in the seabed but continues its forward flow, usually with heavy crevassing on the surface. Ice rumples should be distinguished from *ice rises*, which are areas where ice shelves ground on shoals and the flow is diverted around the shoal. Snow accumulating in the area of the shoal then causes an independent radial flow to be established forming an ice rise. Both rumples and rises act as pinning points in the ice shelf and help to stabilize the ice shelf upstream. This in turn exerts 'back stress' on the inland ice, stabilizing the ice sheet itself. Since the advent of high-resolution imagery, considerable numbers of previously unknown ice rises and rumples have been discovered.

The crests of ice rises are often sharp features, easily identifiable on high-resolution images (Martin and Sanderson, 1980). Unexpectedly, a number of ice rises show a double crest on satellite imagery. An optically levelled line over the ridge of Fletcher Promontory, Antarctica, revealed no topographic origin for the double ridge easily visible on LANDSAT images (J.L.W. Walton, personal communication), thus fuelling speculation about its origin.

(6) *Icefronts* mark the edge of the ice sheet or ice shelf. They advance continuously by ice flow, but retreat infrequently as icebergs calve away from the icefront. Filchner Ice Shelf calved off 11 500 km^2 of ice shelf in a single event, between January and November 1986 (Ferrigno and Gould, 1987). This was the equivalent of 35 years of icefront advance. Monitoring the overall balance of such an unsteady system is

obviously difficult without a precise knowledge of the processes that initiate calving.

A different type of icefront behaviour was recently observed by Doake and Vaughan (1991). The Wordie Ice Shelf was a small (70 × 40 km) ice shelf lying off the west coast of the Antarctic Peninsula. Being the most northerly ice shelf of any significant size on this coast, it was assumed to lie at the climatic limit for ice shelf viability (Mercer, 1978). Comparing the position of the ice front mapped in 1966 from aerial photography with the 1989 icefront position shown by LANDSAT TM, Doake and Vaughan estimated that the ice shelf area has decreased in area from approximately 2000 km^2 to approximately 700 km^2. In fact, the once homogeneous ice shelf had disintegrated into a number of discontinuous ice tongues. Considerable insight into the mechanism of the retreat was gained from the study of successive LANDSAT images taken in 1974, 1979 and 1989. Fracture, either in the form of surface crevasses or rifts extending to the bottom of the ice shelf, has been responsible for an increased iceberg calving rate and weakening of the ice shelf. Fracture appears to have been enhanced by increased amounts of melt water, resulting from a warming trend recorded in mean annual air temperatures in Marguerite Bay. It is predicted that if the current warming trend continues, other nearby ice shelves in the Antarctic Peninsula may be at risk. However, substantial additional warming would be required before similar processes could initiate breakup of the Ross or Ronne-Filchner ice shelves that are believed to stabilize most of the West Antarctic Ice Sheet but at present lie further south than the climatic limit for ice shelves. These ice shelves are probably more prone to shrinkage via increased melting from alterations in ocean circulation.

(7)   *Nunataks* are rock outcrops that protrude through the ice cover. On many images they provide the only fixed points, within a moving sea of ice, by which an image can be rectified. Many areas of ice sheet, however, have no nunataks and so there are many hundreds of kilometres containing no fixed points which have to be bridged before accurate positions can be assigned to features. A method for bridging these areas described by Sievers *et al.* (1990) involves 'tying' many images together using ice features common to overlapping images. A single block adjustment is then performed for all the images of the mosaic, a process that requires significant computer power.

(8)   *Blue ice areas*, where the snow cover is removed by sublimation and wind scouring, occur close to nunataks in many parts of Antarctica. Since the ice surface is periodically removed, meteorites come to the surface and then remain visible for indefinite periods, providing rich pickings for the meteorite hunter (Williams *et al*, 1983). Blue ice areas are also proving to be of great value as landing strips for aircraft not equipped with skis, since the surfaces are strong enough to take a high loading, and the 'sun cupped' surface provides sufficient friction for braking.

Augmentation of visible imagery

Satellite radar altimetry

In theory, satellite radar altimeters should be capable of measuring ice surface elevation to 10 cm precision. Since this is comparable to the accumulation rate of the ice sheet, it ought to be possible to use satellite altimetry to measure any gross imbalance between accumulation and wastage of the ice sheet surface over decadal timescales. On its own, however, radar altimetry does not give a good overall description of the flow regime, nor does it allow easy identification of flow features. Only studies combining visible imagery and radar altimetry can be hoped to give any sound interpretation of ice surface elevation changes.

Satellite radar altimeters have already proved highly effective for obtaining ice sheet topography, despite the fact that none of the instruments flown so far were optimized for use over ice. SEASAT operated for three months in 1978, and GEOSAT from 1985 to 1990. During these missions data was collected over the ice sheets of Greenland and Antarctica up to the orbital limit of 72°. ERS-1, a European satellite launched in July 1991, has an altimeter with a tracking mode specifically designed for use over ice sheets. With an orbital limit of 81° it will also give greatly improved coverage.

There are significant problems to be overcome before satellite radar altimetry can be relied upon for mapping of ice surfaces at 10 cm precision. Electromagnetic penetration of ice at radar frequencies is poorly understood but likely to be of the order of 10 to 100 cm and depends heavily on snow surface conditions (Ridley and Partington, 1988). The individual returned waveforms from ice sheets are highly variable, and both volume scattering and surface scattering are important. It is unclear what point on the waveform corresponds to the ice surface and a number of re-tracking schemes have been proposed. A number of institutes plan fieldwork to perform calibration/validation studies. Over uneven topography the point on the surface giving rise to the radar return, i.e. the point geometrically closest to the satellite, will not generally coincide with nadir. The problem of reconstructing the surface is similar to the migration of three-dimensional seismic data and will require a major effort.

Zwally (1989) and Zwally *et al.* (1989) analysed data from the Seasat and GEOSAT altimeters and found the Greenland Ice Sheet to be thickening by around 0.23 m per year. Although this extensive analysis was performed with both care and thought, the results are difficult to interpret and are the subject of some debate.

Passive microwave

Passive microwave sensors provide brightness temperatures that in general correlate with surface temperatures. As the first microwave images of the polar ice sheets were returned by the Nimbus 5 ESMR it became clear that the correlation is very poor over ice sheets (Chang *et al.*, 1976). More recent studies have shown that microwave emissivity is related to accumulation rates, mean annual temperature and melting effects. Despite the development of a number of models of snow surface emissivity the interpretation of passive microwave data over ice sheets is still a very complicated process. Passive microwave images can only properly be interpreted where a good knowledge of the surface features is already available. Data from the Defense Meteorological Satellite Program (DMSP) SSM/I sensor launched in June

1987, also show promise in glaciological applications, although the resolution of passive microwave imagery is limited to around 30 km by the size of antenna that can be deployed on a satellite at realistic cost.

Synthetic aperture radar

Synthetic aperture radar (SAR) is an active microwave system with both high-resolution and all-weather capability. The SAR system can increase the effective antenna size and hence resolution by the simultaneous processing of waveforms collected along a section of satellite track. Bindschadler *et al.* (1987) showed various images taken by the SEASAT SAR in 1978. Many ice sheet features are visible on these images: undulations, flow lines, crevasses, icebergs, streams and lakes. Since SAR is sensitive to the dielectric variations as well as roughness and surface slope, Bindschadler postulated that much of the detail in SAR imagery is due to variations in the surface composition. Because SAR is a side-looking system, there are inherent distortions in the images known as 'layover', which are difficult to remove without a previously determined terrain model, rarely available over poorly mapped ice sheets. Bindschadler recommended SAR imagery be used in conjunction with visible imagery as together they will constitute a more powerful tool for research.

Conclusions

The use of visible imagery has already revolutionized the study of ice sheet dynamics. Swithinbank's comment on the sorry state of mapping need no longer be applicable. The whole of the Earth's surface is finally accessible without resort to overland survey, for those researchers with the funds to purchase and the time to process and interpret the available imagery.

Despite the tremendous advances in other types of satellite sensors over the past decade, visible imagery will remain the most valuable source of data on the structure and dynamic behaviour of the polar ice sheets. The complete potential of satellite data will, however, only be achieved by integrated interpretation of data from the complete armoury of sensors, together with ground truth wherever it is available.

Acknowledgements

Our thanks are due to all our colleagues at the British Antarctic Survey, especially Dr D.J. Drewry and Dr E.M. Morris, for helpful suggestions and comments.

References

Bindschadler, R.A. and Scambos, T.A., 1991, Satellite-image-derived velocity field of an Antarctic ice stream, *Science*, **252**, 242-6.

Bindschadler, R.A. and Vornberger, P.L., 1990, AVHRR imagery reveals Antarctic ice dynamics, *EOS*, **71**, 23.

Bindschadler, R.A., Jezek, J.C. and Crawford, J., 1987, Glaciological investigations using the synthetic aperture radar imaging system, *Annals of Glaciology*, **9**, 11-19.

Boulton G.S. and Clark, C.D., 1990, The Laurentide Ice Sheet through the last glacial cycle: the topology of drift lineations as a key to the dynamic behaviour of former ice sheets, *Transactions of the Royal Society of Edinburgh: Earth Sciences*, **81**, 327-47.

Casassa, G., Jezek, K.C., Turner, J. and Whillans, I.M., 1991, Relic flow stripes on the Ross Ice Shelf, *Annals of Glaciology* (in press).

Chang, A.T.C., Gloersen, P., Schmugge, T., Wilheit, T.T. and Zwally, H.J., 1976, Microwave emission from snow and glacier ice, *Journal of Glaciology*, **16**, 23-39.

Choudry, B.J. and Chang, A.T., 1981, On the angular reflectance of snow, *Journal of Geophysical Research*, **86 C1**, 465-72.

Crabtree, R.D. and Doake, C.S.M., 1980, Flow lines on Antarctic ice shelves, *Polar Record*, **20**, 31-7.

Doake, C.S.M. and Vaughan, D.G., 1991, Rapid disintegration of Wordie Ice Shelf in response to atmospheric warming, *Nature*, **350**, 6316, 328-30.

Dowdeswell, A.J. and McIntyre, N.F., 1986, The saturation of LANDSAT images over large ice masses, *International Journal of Remote Sensing*, **7**, 151-64

Ferrigno, J.G. and Gould, W.G., 1987, Substantial changes in the coastlines of Antarctica revealed by satellite imagery, *Polar Record*, **23 (146)**, 577-83.

Ferrigno, J.G. and Molnia, B.F., 1989, Availability of LANDSAT, Soyuzkarta and SPOT data of Antarctica for ice and climate research, *Antarctic Journal of the United States*, **24**, 15-18.

Frolich, R.M., Vaughan, D.G. and Doake, C.S.D., 1989, Flow of Rutford Ice Stream and comparison with Carlson Inlet, Antarctica, *Annals of Glaciology*, **12**, 51-6.

Gessel, G., 1989, An analysis for snow and ice detection using AVHRR data. An extension of the Apollo software package, *International Journal of Remote Sensing*, **10**, 897-905.

Grosfeld, K., Hinze, H., Ritter, R., Shenke, H.W., Sievers, J. and Thyssen, F., 1989, *Ekströmisen, Anarktis 1:500 000. Maps of ice shelf kinematics*. Frankfurt: Institut für Angewandte Geodäsie.

Keys, H.R.J., Jacobs, S.S. and Barnett, D., 1990, The calving and drift of Iceberg B-9 in the Ross Sea, Antarctica, *Antarctic Science*, **2**, 243-59.

Krimmel, R.M. and Meier, M.F., 1975, Glacier applications of ERTS images, *Journal of Glaciology*, **15**, 391-402.

Luchitta, B.K. and Ferguson, H.M., 1986, Antarctica: Measuring glacier velocity from satellite images, *Science*, **28**, 1105-8.

MacAyeal, D.R., Bindschadler, R.A., Jezek, K.C. and Shabtaie, S., 1988, Can relic crevasse plumes on Antarctic ice shelves reveal a history of ice stream fluctuations? *Annals of Glaciology*, **11**, 77-82.

Martin, P.J. and Sanderson, T.J.O., 1980, Morphology and dynamics of ice shelves, *Journal of Glaciology*, **25**, 33-45.

Mercer, J.H., 1978, West Antarctic Ice Sheet and CO_2 greenhouse effect: A threat of disaster, *Nature*, **271**, 321-5.

Orheim, O. and Luchitta, B.K., 1987, Snow and ice studies by thematic mapper and multispectral scanner LANDSAT images, *Annals of Glaciology*, **9**, 109-18.

Østrem, G., 1975, ERTS data in glaciology—An effort to monitor glacier mass balance from satellite imagery, *Journal of Glaciology*, **15**, 403-15.

Ridley, J.K. and Partington, K.C., 1988, A model of the satellite radar altimeter return from ice sheets, *International Journal of Remote Sensing*, **9**, 601-24.

Shabtaie, S. and Bentley, C.R., 1987, West Antarctic ice streams draining in to the Ross Ice Shelf: configuration and mass balance, *Journal of Geophysical Research*, **92** (B2), 1311-36.

Sievers, J., Grindel, A. and Meier, W., 1990, Digital image mapping of Antarctica, *Polarforschung*, **59**, 25-35.

Swithinbank, C.W.M., 1977, Glaciological research in the Antarctic Peninsula, *Philosophical Transactions of the Royal Society of London, Series B*, **279**, 161-83.

Swithinbank, C.M.W. and Luchitta, B.K., 1986, Multispectral digital image mapping of Antarctic ice features, *Annals of Glaciology*, **8**, 159-63.

Swithinbank, C.M.W., McClain, P. and Little, P., 1977, Drift tracks of Antarctic icebergs, *Polar Record*, **18**, 495-501.

Swithinbank, C.M.W., Brunk, K. and Sievers, J., 1988, A glaciological map of Filchner-Ronne Ice Shelf, Antarctica, *Annals of Glaciology*, **11**, 150-6.

Turner, J., 1987, Detection of clouds over ice, in *Satellite and Radar Imagery Interpretation*. Darmstadt, Germany: European Organisation for the Exploitation of Meteorological Satellites (EUMETSAT), 521-39.

Vaughan, D.G., Doake, C.S.M. and Mantripp, D.R., 1988, Topography of an Antarctic ice stream, in *SPOT 1 Image Utilization, Assessment, Results*, Toulouse, France: CNES. Cepadues-Editions, 167-74.

Warrick, R. and Oerlemans, J., 1990, Sea level rise, in Houghton, J.T., Jenkins, G.J. and Ephraums, J.J. (Eds) *Inter-Governmental Panel on Climate Change: Working Group 1— Climate Change: The IPCC Scientific Assessment*, Cambridge: Cambridge University Press.

Wildey, R.L., 1984, Topography from single radar images, *Science*, **224**, 153-6.

Williams, R.S., Meunier, T.K. and Ferrigno, J.G., 1983, Blue ice, meteorites and satellite imagery in Antarctica, *Polar Record*, **21**, 493-504.

Zwally, H.J., 1989, Growth of the Greenland Ice Sheet: interpretation, *Science*, **246**, 1589-91.

Zwally, H.J., Brenner, A.C., Major, J., Bindschadler, R.A. and Marsh, J.G., 1989, Growth of the Greenland Ice Sheet: measurement. *Science*, **246**, 1587-88.

Chapter 3
'Long-term' land surface processes: erosion, tectonics and climate history in mountain belts

B.L. ISACKS

Institute for the Study of the Continents,
Department of Geological Sciences,
Cornell University, Ithaca, New York

Introduction

In studies of global climate change induced by human activities, 'land surface processes' refer to relatively short-term diurnal to decadal interactions of the atmosphere with the land surface biosphere and hydrosphere. These interactions operate in a framework of geological processes including weathering, soil formation, erosion, sediment transport and accumulation, uplift, crustal deformation, deep crustal fluid flow, and magmatism. The geological processes are generally assumed to be so slow in action as to form a kind of static, non-interactive framework for the more rapidly acting processes of the hydrosphere and biosphere. Although this may be so to some extent in low relief forests, grasslands, and agricultural or industrial regions, geological processes play a prominent if not leading role in the interactions of the land surface in mountainous regions of earth. These interactions cover a broad range of timescales from those of concern to human adaptation to the evolutionary changes in the Earth system over geological eras.

Tectonic uplift and magmatism produce mountains which have significant orographic affect on both regional and global climate. Although the regional effects have been long recognized, only recently have the profound global effects of continental scale mountain belts become appreciated. In turn, the climate affects the evolution of the mountain belt through the chemical and mechanical weathering of uplifted rocks and the rapid erosional transfer of crustal mass away from the mountain belt. The mountain landscape evolves as the product of tectonic construction interacting with climatically driven erosional processes (Figure 3.1). Complex feedback loops are possible because the form of the uplifting mountain mass so strongly affects regional climate and vegetation, while the erosional transfer of crustal mass affects the tectonic stress system in the crust and mantle that produces the uplift.

The 'geological' timescales for the mountain/atmosphere interaction, of the order of 10^7 years, concern the thermal, deformational, and magmatic

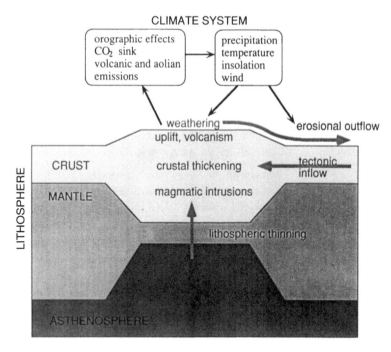

Figure 3.1   Schematic cross sectional view of tectonic and climate interactions in a continental mountain belt. The uplift is produced by addition of light crustal mass from the mantle (magmatism) and by lateral shortening and thickening (tectonic inflow), and by lithospheric thinning (the boundary between lithosphere and asthenosphere is an isotherm). Outflow of crustal mass from the system by erosion is determined by climate, while the uplift, volcanism, and wind erosion (dust) affect the atmosphere and climate.

processes producing the continental-scale uplift of a mountain belt. The uplift can be associated with similar scale temporal variations in climate. For example, Ruddiman *et al.* (1989) argue that the late Cenozoic uplift of mountains in Asia and North America was a major factor in Late Cenozoic global cooling that culminated in the cycles of major continental glaciations of the past several million years.

At the other end of the spectrum are the very short duration 'events', such as earthquakes, landslides, glacial surges, severe storms and floods, which have durations of minutes to months. The roles of rare, short duration, but very high energy events in hydrological, erosional and tectonic processes are quite similar. In each case the sizes or magnitudes of the events constitute a continuum characterized by a power law relationship between frequency of occurrence and magnitude. The logarithm of the number of events per unit time decreases linearly with the logarithm of magnitude over a large range of sizes, although it is generally supposed that these fractal-like distributions have some physically determined upper limit. Since the larger, more infrequent events do most of the work, their sizes and recurrence intervals limit the rate of action of the process. Understanding the role of the large, rare events requires continental- to global-scale observations to improve the chances of 'catching' an event. Study of Earth history is necessary to estimate rates of occurrence, determine the cumulative effects of the events, and to identify events that recur beyond the scale of human observation.

This paper focuses on erosion as a key part of the mountain/atmosphere system and considers the following thesis: erosion rates, although often averaged over large areas and time spans, are actually characterized by a high degree of concentration in both space and time: in space, within certain narrow belts determined by the interaction of tectonics and climate, and in time, during particular phases of the Quaternary glacial cycles. This natural heterogeneity in time and space is the framework within which to consider the dramatic increase in erosion rates caused by human activities.

The combination of high rates of uplift driven by active tectonics and a strong interaction with climate is found most extensively in the Himalayas and adjacent regions of central Asia and in the equatorial and sub-tropical Andes. In these areas large, continental-scale plateaus are being eroded primarily along the edges upon which a strong moisture flux impinges. The moisture flux in the Andes is associated with the Atlantic-Amazonian Inter-Tropical Convergence Zone, while that in the Himalayan regions is associated with the Asiatic monsoon. Smaller regions of strong tectonic/climate interaction include areas such as the Southern Alps of New Zealand, the New Guinea Highlands and Taiwan. In all these regions, crustal deformation, isostasy and orographic precipitation produce narrow belts of high topographic relief with probably the world's highest erosion rates. Furthermore, these rates are likely to have been amplified considerably during particular phases of the glacial-interglacial climate cycle.

The Cornell EOS Interdisciplinary Science team[†] is exploring these ideas in a comprehensive study of the Andes Mountains. The object of the project is to study spatial and temporal variations in climate, erosion and tectonics on the scale of a major continental mountain belt. The scale of the study is possible only with the broad spatial and temporal coverage of key topographic, geological and climate-related land surface properties provided by satellite instruments. The Cornell team seek to determine the presently acting tectonic and erosional regime, its spatial variability and rates of change, and to compare the present erosional and climatic regime with that during the last glacial maximum. The effects of the last glacial maximum are extensively recorded in the Andean landscape by glacial cirques, moraines, paleo-lake shores and wind direction indicators. The comprehensiveness, regional scale and integration of this EOS study has no precedent in traditional geomorphology or climatology. The project is made possible by

†B.L. Isacks (Principal Investigator), R.W. Allmendinger, A.L. Bloom, E.J. Fielding, T. Jordan, S.M. Kay and W. Philpot.

availability of satellite data, together with the accelerated evolution of
computer capabilites during the past decade.

Erosion rates

Erosion or denudation rates, the rates at which mass is removed from the
land surface (volume removed per unit area per unit time or thickness
removed per unit time) have been most often estimated from measurements
of the mass of dissolved and suspended material carried by a river exiting a
drainage basin over some period of time. The erosion rate is then obtained by
converting to volume and dividing the volume of material passing through
the river per unit time by the area of the drainage basin to obtain an average
thickness of rock removed from the basin per unit time. A convenient
compilation of such estimates from Pinet and Souriau (1987) is shown in
Figure 3.2 for large drainage basins. The rates obtained by this approach give
no indication of the spatially concentrated zones of very high rates that
clearly exist within many of the basins. For example, the low rate for the
Amazon basin (0.06 mm/yr) is controlled by the enormous area of low relief

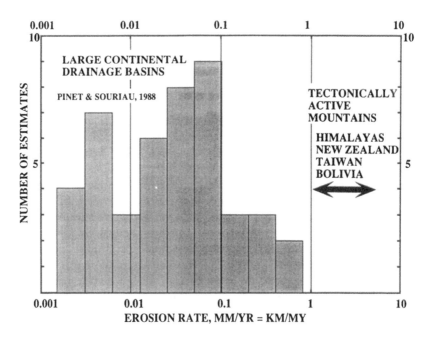

*Figure 3.2   Histogram of erosion rates reported by Pinet and Souriau (1987). The high erosion
rates shown on the right side of the figure are reported by Zeitler (1985) and Hubbard et al.
(1991) for the Himalayas; Adams (1980) and Whitehouse (1986) for New Zealand; Li (1975)
for Taiwan; and Benjamin et al. (1987) for Bolivia.*

and low erosion rate drained by the river and does not reveal the very high rates of erosion in the Andean headwaters region. Storage of sediment within a basin may also decrease the apparent erosion rate averaged over an entire basin. The foreland basins east of the Andes hold large thicknesses of sediment eroded from the mountains.

In contrast, the highest rate compiled by Pinet and Souriau (1987) is for the Yellow River basin of China. This rate is primarily a result of the large amount of material supplied from the highly-erodable loess belt crossed by the river. This indicates one reason for a very high erosion rate, a high degree of erodability of the material. In this paper we focus on the high rates produced by the tectonic effect on uplift and climate.

As Stallard (1988) points out, the highest erosion rates characteristically occur in tectonically active regions. Some of the highest estimates are illustrated in Figure 3.2. Three key factors combine in these regions to produce the high erosion rates: (1) high relief provides abundant gravitational potential energy, (2) young and generally more erodable rocks are exposed, and (3) orographic effects on atmospheric circulation concentrate precipitation and maximize weathering rates and transport capability. In such regions erosion rates are weathering rather than transport limited (Carson and Kirkby, 1972). The landscape is characterized by steep straight slopes and high relief developed by rapidly acting processes of hillslope development such as landsliding and rock avalanching together with actively downcutting drainage systems. Soils are generally thin and youthful.

Tectonic controls

Where the orographic effect concentrates precipitation and produces high rates of erosion, the tectonic interaction can serve to maintain these high rates by feeding both gravitational energy and crustal mass into the system, this leading to large quantities of crustal material being processed through the weathering–erosional part of the rock cycle. This is accomplished partly by isostasy, which represents stored gravitational energy produced by the tectonic forces that thickened the crust in the mountain belt. We discuss the simplest isostatic effect below, then in the following section we look at some specific examples of the tectonic effects in the Himalayas and Andes.

Simple isostatic effect

An extremely simplified, one dimensional model of the isostatic–erosional effect is shown in Figure 3.3. In this simple model without relief, mass is uniformly eroded from the column with instantaneous isostatic compensation. Uplift of the column is produced by inflow of mantle (asthenospheric) mass so that pressure at some depth of compensation remains constant. If the erosion rate is proportional to the elevation of the top of the column above a

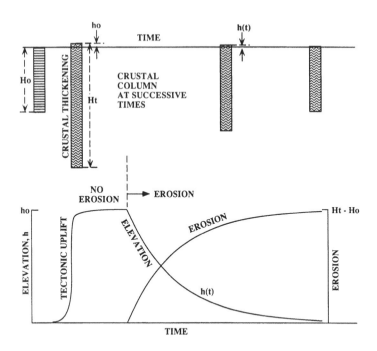

Figure 3.3   Simple one-dimensional model of isostatic effect of erosion where crustal thickening and uplift occur first, then erosion (uniform over the column) proceeds without further tectonic input. The erosion rate is taken to be proportional to the elevation above a base level, and the isostatic response is assumed to be instantaneous. In the upper figure a crustal column is shown at several times with respect to a stable baseline (the horizontal line). The total crustal thickness before thickening is H_o and the thickness after crustal thickening and uplift is H_t. Elevation, $h(t)$ has the value h_o after crustal thickening but before erosion starts, then decays exponentially with time as shown in the lower plot. The thickness of material eroded from the column is shown as 'erosion' and increases from 0 to the amount of crustal thickening, H_o-H_t. Note the large difference between the right-hand scale for erosion and the left-hand scale for elevation; For $h_o=4$ km, H_t-H_o will be about 32 km.

base level, an assumption often used (e.g. Ahnert, 1970; Pineau and Soiret, 1987), then the exponentially-decaying response shown in Figure 3.3 is produced. Time constants of the orders of 10^6 to 10^8 years are cited for lowering the average elevations of entire drainage basins. Though highly oversimplified, the model shows how isostasy feeds crustal material into the erosional processor. The model also clearly distinguishes the various types of vertical motions involved. Average elevation above sea level exponentially decays from its initial value of only several kilometres, while thickness of the column eroded can reach several tens of kilometres corresponding to the initial excess crustal thickness.

Real topographic relief, with its fractal characteristics spanning an enormous range of spatial scales and its hierarchical drainage network, combine with the complex spatial variations in erodability and precipitation to make the problem a great deal more complex. Computer models of eroding landscapes are only just beginning to emerge (e.g. Willgoose *et al.*, 1990). The conceptual views developed in geomorphological research during this century starting with Davis (1899; see also Bloom, 1991), however, provide a qualitative framework depicted schematically in Figure 3.4.

We consider a representative column with an areal size that is small compared to the overall extent of the mountain belt but large enough to contain a significant representation of local relief and the drainage network.

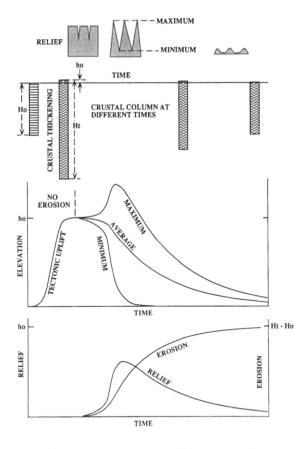

Figure 3.4   The model of Figure 3.3 with erosion of the column producing relief schematically illustrated in the top figures. Again, note the large differences in the right- and left-hand scales of the lower plot. The figure is constructed assuming that the minimum elevations, controlled by the downcutting drainage system, decrease with a shorter time constant than that for the uplands. The actual shape of such curves depends upon the evolving forms of the landscape resulting from the interaction of climate, lithology and structure, and remain major unsolved research problems.

Here local relief is given by the difference between maximum and minimum elevations within the defined spatial window. The maximum and minimum elevations within this window can be plotted as the column erodes. The form of the developing relief determines the relationship between the average elevation and the minimum and maximum elevations. For example, in the scheme of Figure 3.4 the initial stages of relief are characterized by V-shaped canyons cutting into the uplifted surface. As erosion begins, headward development and downcutting of rivers proceed rapidly while much of the uplifted upland region remains little eroded. As the fluvial system and steep-sided valley slopes include all of the area, the average slope of the region reaches a maximum. It is reasonable to suppose that during this stage the erosion rates are a maximum, with gravity driven mass movements such as landslides predominating on the steep slopes and a vigorous fluvial system downcutting the column and rapidly flushing the eroded mass out. The minimum elevations would correspond to the highest order drainage in the column.

Isostasy causes the maximum elevations in the uplands to actually rise during the initial period. The rapid downcutting of the canyons relative to the uplands lowers the average elevation, while the associated isostatic response uplifts the less eroded highland areas. As mass from both peaks and valleys becomes efficiently removed, a stage of maximum erosion rate occurs, characterized by high relief, rising peaks and an actively downcutting drainage system. The system is weathering limited, with more resistant lithologies—generally the more metamorphosed rocks—forming the higher massifs. Eventually, in the absence of renewed tectonic uplift, the entire envelope of elevations decreases. As the minimum elevations approach a stable base level, average slopes and erosion rates in the column area decrease.

Tectonic-erosional mass budget

In addition to the isostatic effect, regional tectonic deformations and magmatism may replenish crustal mass lost by erosion, maintain the phase of high average elevation, high relief and high erosion rates shown in Figure 3.4, leading to the processing of large quantities of crustal material. There are three possible situations for the crustal tectonic–erosional mass budget:

(1) uplift driven by tectonics where erosion is unable to keep up with the increases in crustal mass;
(2) a steady state where the tectonic input is balanced by the erosional output; and
(3) the tectonic input which is too slow relative to erosion, so the erosional/isostatic response proceeds to lower and smooth the landscape.

Figure 3.5 schematically portrays the first two cases.

The best cases for a steady state balance are made by Adams (1980) for the New Zealand Southern Alps and Suppe (1981) for the Taiwanese collision. Taiwan is a rather special case: nearly the entire mountain belt and adjacent fold thrust belt involves Late Cenozoic sediments scraped up and deformed by the collision of an island arc with a continental margin. The deformation in this case is remarkably well modelled as a propagating Coulomb wedge (Dahlen and Suppe, 1988), with erosion at the top (rear) of the wedge compensating for input of material at the toe of the wedge.

The arid regions of the central Andes and Tibet offer clear illustrations of the first case (top diagram in Figure 3.5), where erosion rates are low relative to tectonically driven uplift and constructional volcanism. The plateau uplift of the central Andean plateau is accommodated by a crustal scale flexure on the western side that is veneered by a Late Cenozoic volcanic cover (Isacks, 1988). The well preserved, largely intact features of Late Cenozoic tectonics and magmatism dominate the character of the landscape in the arid to hyper-arid regions of the western Andes of northern Chile, western Bolivia and northwestern Argentina. Abele (1989) indicates how the Andean uplift itself has helped preserve and strengthen the aridity of this region through the continental-scale orographic effect on southern hemisphere atmospheric circulation.

Effects of crustal deformation

The most dynamic systems in which the erosional and crustal mass fluxes are both high require consideration of how the geometry of crustal deformation produces uplift and the ensuing climate interaction. A most interesting possibility is that the systems are closely coupled in the following sense: the tectonic deformation leads to an orographic effect which amplifies the

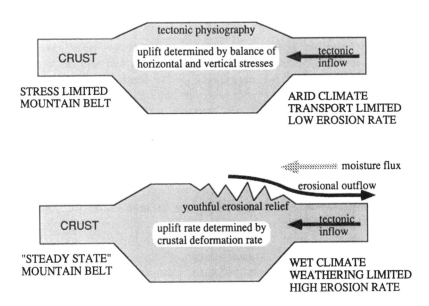

Figure 3.5   Schematic diagram of two extreme cases of the tectonic–erosional crustal mass budget.

erosional flux in certain parts of the system; the removal of mass in those areas then feeds back into the tectonic stress system in such a way as to enhance the crustal deformation producing the orographic effect, thereby maintaining or increasing the orographic effect and exposing new crustal material to be eroded.

The crustal deformational geometries involved in most active regions of compressional mountain building involve crustal shortening and thickening associated with crustal-scale thrust faulting. The specific geometries and development of these systems remain subjects of controversy and probably differ significantly from region to region. Models for Taiwan and the New Zealand Southern Alps are presented by Suppe (1981, 1987) and Koons (1990), respectively, based on the idea of crustal material piling up as if scraped up by a giant snow plow—as a propagating, self-similar Coulomb wedge. Zhao and Morgan (1985) and Isacks (1988) present models for the Himalayas and the Andes, respectively, based on a two-fold division of the crustal rheology into a relatively strong brittle upper crust and a ductile lower crust. However, the deep crustal geometry of deformation remains a major unsolved research problem. We illustrate below the role of tectonics in the Andes, the focus of our EOS study. We also show a comparison with the Himalayan mountain belt, which, though quite different in overall size, tectonic setting, and amount of uplift, exhibits some very interesting similarities to the Andes with respect to the relationships of topography, climate and tectonics.

Eroding plateau edges in the Andes and the Himalayas

The world's highest and most extensive plateaus are, first, the Tibetan plateau at an average elevation of near 5 km over a vast area of central Asia and, second, the Altiplano-Puna plateau of the central Andes with average elevations near 4 km over a width of about 300 km extending from latitudes 12°S to 28°S. Both uplifts result from convergence of two lithospheric plates: the Tibetan plateau from collision of continental plates, and the Andean plateau from subduction of sub-oceanic lithosphere beneath a continental plate. Both plateaus are formed primarily by horizontal shortening and vertical thickening of the crust. The interior parts of the two plateaus are internally drained and relatively arid. In both cases the main surface expressions of active crustal shortening are located along marginal thrust belts, the Himalayan thrust system along the southern edge of the Tibetan plateau and the Eastern Cordilleran/Subandean thrust system along the eastern edge of the central Andean plateau.

The eastern Andean thrust belt extends south from equatorial latitudes to the mid-latitude temperate zone, and has a climate variation ranging from tropical rain forest to semi-arid to arid desert along this long latitudinal span. Figure 3.6 shows high relief areas of the central Andes derived from a digital elevation model (Isacks, 1988) for the Andean region. The relief shown refers to the maximum–minus–minimum elevations determined for overlapping windows of about 25 km in dimension. The areas of high relief shown in the figure reflect mainly fluvially and glacially dissected terrain or in certain regions short-wavelength volcanic or tectonic topography. The very high erosional relief along the northeastern edge of the Altiplano is associated with erosional effects of the strong moisture fluxes from the Amazon basin. The high relief region near 31°S–34°S is associated with the southwardly increasing moisture flux from the mid-latitude westerly winds.

Many of the high relief areas shown in Figure 3.6 are also associated with areas which were most heavily glaciated during the last glacial maximum (for examples see Clapperton, 1983). One of the primary findings of our ongoing study of Andean climate during the last glacial maximum is that the main change from present climate is intensification of modern patterns rather than major shifts of atmospheric circulation patterns. The implications of this intensification for higher erosion rates in the Pleistocene is discussed in the next section.

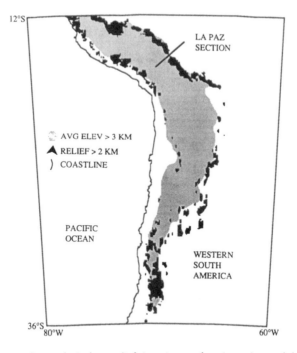

Figure 3.6 Map of central Andean relief (maximum elevation minus minimum elevation) determined within a moving spatial window of 0.25° latitude by 0.25° longitude applied to the digital elevation model of Isacks (1988). Only areas with relief above 2 km are shown. The shaded region of average elevations greater than 3 km is determined from the same database with the same spatial window.

The fluvially and glacially dissected region of high relief along the northeastern edge of the Altiplano indicates a narrow belt of high erosion rate produced by the orographic focusing of precipitation on an active thrust front. Strong moisture fluxes from the Amazon basin impinge on the plateau from the northeast and develop a high precipitation zone along the steep northeastern slopes of the Altiplano plateau and a corresponding precipitation shadow within the plateau itself.

Figure 3.7 illustrates the characteristics of this region in a cross section through the northeastern edge of the Bolivian Altiplano near La Paz. The cross section is taken normal to the two-dimensional, long-wavelength form of the plateau edge. Digital topography is compiled within a 50 km wide zone or swath along the cross section. The profile shows maximum, minimum, and averaged elevations within a moving window that is 50 km wide perpendicular to the cross section and 10 km in dimension along the section. The relief is determined as the maximum minus the minimum

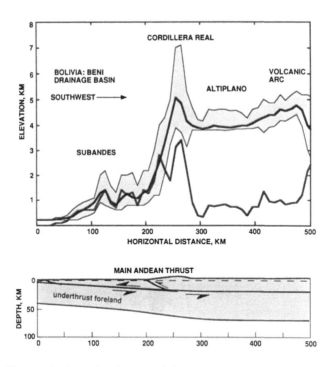

Figure 3.7   Topographic (upper) and structural (lower) cross sections in the region of La Paz, Bolivia. The trace of the section is shown on the map of Figure 3.6. The maximum (upper curve bounding the shaded region), minimum (lower curve bounding shaded region) and average (thick curve within shaded region) elevations and relief (lighter thick curve) are computed within spatial windows 10 km along the section line and 50 km on either side of the section line. Note the large amount of vertical exaggeration indicated by the horizontal and vertical scales. The lower section shows schematically the thrust geometries within the crust (shaded area) with no vertical exaggeration. The dashed horizontal line represents the sea level.

elevation for each of the successive spatial windows. This representation of relief can be interpreted in terms of the concepts illustrated in Figure 3.4 with the spatial variation of precipitation and erosion due to the orographic effect of the windward facing plateau edge replacing the temporal parameter in the one-dimensional model of Figure 3.4. As the orographic precipitation/ erosion front moves into the plateau, an initially uplifted but un-eroded section on the arid plateau can be thought to pass through the stages indicated in Figure 3.4. Thus, in an approximate way, the zone of maximum relief shown in Figure 3.7 indicates the band of high erosion rates that is associated with the orographically controlled zone of high precipitation and the tectonically controlled steep slope of the plateau edge. The central problem in understanding the interaction of climate and tectonics here is the role of active tectonics in feeding material to the edge.

In this respect an interesting comparison can be made between the Bolivian plateau edge and the southern edge of the Tibetan plateau formed by the Himalayan mountains. Figure 3.8 shows a topographic cross section

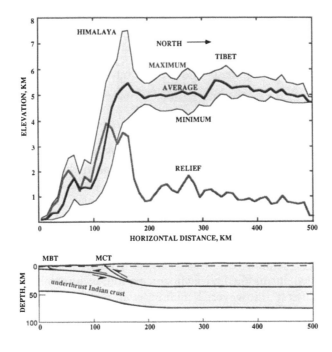

Figure 3.8   Cross section through the Himalayas in the region of Nepal near Katmandu using the same symbolism as Figure 3.7. The cross section is taken perpendicular to the local WNW–ESE strike of the Himalayas. The database includes digital elevations specified on a 100 m grid. 'MBT' and 'MCT' refer to the Main Boundary Thrust and Main Central Thrust often used in discussions of Himalayan tectonics. The structures are simplified from Ni and Barazangi (1984).

constructed in a similar fashion to Figure 3.7. The Himalayan belt is subjected to the large moisture fluxes associated with the Asiatic monsoon (impinging from left to right in Figure 3.8). While the areal extent and elevation of the Tibetan plateau are greater than the Altiplano, the profiles in Figures 3.7 and 3.8 show that the two plateau edges have a remarkably similar shape and relief profile.

The two plateau edges share an analogous tectonic control as illustrated in the lower sections of Figures 3.7 and 3.8. Below the topographic profiles are simplified structural sections according to Roeder (1988) for the Bolivian section and Ni and Barazangi (1984) for the Himalayan section. Both show crustal-scale 'ramp-flat' thrust structures which bring deep crustal material to the surface. The sections show the intimate relationships between the regions of high relief, high overall slope and the thrust systems. In each case the thrust system separates a shallow thin-skinned fold-thrust belt from the steep edge of the high plateau where more pervasively deformed, magmatically intruded, older and more deeply formed rocks are exposed. The thrust zones move crustal material upwards into the orographically controlled belt of high precipitation, and thereby tend to maintain the high rates of erosion and high relief. The erosion can directly affect the rate of thrusting by unloading and thereby lessening the work done by the thrusting against gravity. The extent to which the erosional unloading affects the rate and geometry of thrusting is still to be modelled and explored by comparing thrusting rates in regions with different climates and erosion rates.

Quaternary amplification of erosion rates

Both Roeder (1988) and Benjamin et al. (1987) estimate Late Cenozoic removal of a thickness of 10–20 km of material in the area of high relief northeast of La Paz (depicted in Figure 3.7). This yields overall average rates of about 1 mm/yr integrated over the Neogene. Isacks (1988) finds a deficit of topographic mass in this region in terms of a model of crustal shortening that is compatible with the development of the Bolivian 'orocline' (the bend in the shape of western South America in the region of the Bolivia and southern Peru). This deficit is also largely accounted for by the mass of material eroded during the Late Cenozoic period, in agreement with Roeder and Benjamin et al. However, Benjamin et al. (1987) interpret their fission track data to yield rates of mass removal as high as 5 mm/yr during Quaternary times. Similarly high rates are reported for parts of the Himalayas.

The effects of glaciation increasing erosion rates in high relief regions is well known. These effects include the scouring, widening and deepening of valleys, the steepening of valley sides and the storage of water in the high elevation glaciers which, when released during deglaciation, flushes out the large amounts of glacially excavated material. It is likely, therefore, that regions more heavily glaciated during the colder periods of the Quaternary climate cycles had high erosion rates during relatively short periods of time, possibly corresponding to the deglaciation phases of the cycles when large amounts of melt water are flushed through the fluvial system. The overall Late Cenozoic average rates for such regions would be dominated by such phases of intense erosion. This can be considered an extension of the concept that large rare 'events' do most of the geomorphic work.

An important aspect of the system shown in Figure 3.1 is the role of weathering of silicate minerals as a sink for atmospheric CO_2 (Berner et al., 1983). Raymo et al. (1988) and Molnar and England (1990) point out the positive feedback mechanisms: increased erosion rate leads to reduction in atmospheric CO_2, which then increases the erosion rates by increasing the extent of storminess and glaciation. The increased erosion rate then further decreases the concentration of CO_2 in the atmosphere. The timescale proposed for this feedback loop is that of the Late Cenozoic global cooling. However, if the very high erosion rates in active mountain belts have characteristic durations comparable to a scale of 10^2 to 10^4 years involved in the Quaternary climate oscillations, then the geological feedback effect in the global carbon cycle may be important even on these short timescales.

Conclusions

The brief discussion presented here on the temporal and spatial scales of high 'natural' erosion rates in continental regions points to the energetic interactions of the atmosphere, tectonics and land surface hydrology in mountainous regions of the Earth. These interactions are now just beginning to be studied with the full panoply of satellite observations and other massive spatial databases such as the digital elevation models illustrated in this presentation. These vast new data sets provide synoptic views of continental-scale regions in unprecedently comprehensive and thorough ways. They are promoting the integration of the hitherto compartmentalized disciplines of climatology, hydrology, structural geology, geomorphology and climatology into a unified study of physical land surface processes.

References

Abele, G., 1989, The interdependence of elevation, relief, and climate on the western slope of the Central Andes. *Zbl. Geol. Palaont., Teil I,* (5/6), 1127–39.

Adams, J., 1980, Contemporary uplift and erosion of the Southern Alps, New Zealand, *Bulletin of the Geological Society of America,* **91**, II, 1–114.

Ahnert, F., 1970, Functional relationships between denudation, relief and uplift in large mid-latitude drainage basins, *American Journal of Science,* **268**, 243–63.

Benjamin, M.T., Johnson, N.M. and Naeser, C.W., 1987, Recent rapid uplift in the Bolivian Andes: evidence from fission-track dating, *Geology,* **15**, 680–3.

Berner, R.A., Lasaga, A.C. and Garrels, R.M., 1983, The carbonate-silicate geochemical cycle and its effect on atmospheric carbon dioxide over the past 100 million years, *American Journal of Science*, **283**, 641–83.

Bloom, A., 1991, *Geomorphology: A systematic analysis of late Cenozoic landforms*, Englewood Cliffs, New Jersey: Prentice Hall, 532 pp.

Carson, M.A. and Kirkby, M.J., 1972, *Hillslope form and process*, London: Cambridge University Press, 475 pp.

Clapperton, C.M., 1983, The glaciation of the Andes, *Quaternary Science Reviews*, **2**, 83–155.

Dahlen, F.A. and Suppe, J., 1988, Mechanics, growth and erosion of mountain belts, *Special Paper, Geological Society of America*, **218**, 161–78.

Davis, W.M., 1899, The geographical cycle, *Geographical Journal*, **14(A)**, 481–503.

Hubbard, M., Royden, L. and Hodges, K., 1991, Constraints on unroofing rates in the high Himalaya, Eastern Nepal, *Tectonics*, **10**, 287–98.

Isacks, B.L., 1988, Uplift of the Central Andean Plateau and the bending of the Bolivian Orocline, *Journal of Geophysical Research*, **93**, 3211–31.

Koons, P.O., 1990, Two-sided orogen: collision and erosion from the sandbox to the Southern Alps, New Zealand, *Geology*, **18**, 679–82.

Li, Y.H., 1975, Denudation of Taiwan Island since the Pliocene Epoch, *Geology*, **4**, 105–7.

Molnar, P. and England, P., 1990, Late Cenozoic uplift of mountain ranges and global climate change: chicken or egg? *Nature*, **346**, 29–34.

Ni, J. and Barazangi, M., 1984, Seismotectonics of the Himalayan Collision Zone: geometry of the underthrusting Indian Plate beneath the Himalaya, *Journal of Geophysical Research*, **89**, 1147–63.

Pinet, P. and Souriau, M., 1987, Continental erosion and large-scale relief, *Tectonics*, **7**, 563–82

Raymo, M.E., Ruddiman, W.F. and Froelich, P.N., 1988, Influence of Late Cenozoic mountain building on ocean geochemical cycles, *Geology*, **16**, 649–53.

Roeder, D., 1988, Andean-Age structure of Eastern Cordillera (province of La Paz, Bolivia), *Tectonics*, **7**, 23–39.

Ruddiman, W.F., Prell, W.L. and Raymo, M.E., 1989, Late Cenozoic uplift in Southern Asia and the American West: rationale for general circulation modeling experiments, *Journal of Geophysical Research*, **94**, 18379–91.

Stallard, R.F., 1988, Weathering and Erosion in the Humid Tropics, in Lerman, A. and Meybeck, M. (Eds.), *Physical and Chemical Weathering in Geochemical Cycles*, Dordrecht: Kluwer Academic Publishers, 225–46.

Suppe, J., 1981, Mechanics of mountain building and metamorphism in Taiwan, *Memoirs of the Geological Society of China*, **4**, 67–898.

Suppe, J., 1987, The active Taiwan mountain belt, in Shaer, J.P. and Rodgers, J., *The Anatomy of Mountain Ranges*, Princeton, New Jersey: Princeton University Press, 277–93.

Whitehouse, I.E., 1986, Geomorphology of a compressional plate boundary, Southern Alps, New Zealand, *International Geomorphology*, Part 1, 897–924.

Willgoose, G., Bras, R.L. and Rodriguez-Iturbe, I., 1990, A model of river basin evolution, *EOS, Transactions of the American Geophysical Union*, **71**, 1806.

Zeitler, P.K., 1985, Cooling history of the NW Himalaya, Pakistan, *Tectonics*, **4**, 127–51.

Zhao, W.-L. and Morgan, W.J., 1985, Uplift of the Tibetan Plateau, *Tectonics*, **4**, 359–69.

Chapter 4

Applications of satellite remote sensing techniques to volcanology

P. FRANCIS and C. OPPENHEIMER

Department of Earth Sciences,
The Open University

Introduction

The 1980s were the worst decade for volcanic disasters since the beginning of the 20th century. Over 24 000 people were killed as a result of two eruptions: El Chichón in Mexico (1982) and Nevado del Ruiz in Colombia (1985). Other eruptions in the United States, Italy and Indonesia claimed some 100 lives, and a further 1800 died after carbon dioxide bursts from volcanic crater lakes in Cameroun in 1984 and 1986 (Tilling, 1989). At the time of writing, ongoing eruptions at Unzen in Japan, and Pinatubo in the Philippines, had resulted in many deaths and necessitated large-scale evacuations. It is self-evident that the local, regional, and even global impacts of eruptions demand that potentially active volcanoes be monitored regularly, and that volcano-logists should endeavour to interpret volcano behaviour so that meaningful predictions can be made. 530 volcanoes have erupted in historic times, and a further 2600 have been active in the last 10 000 years (C. Newhall, personal communication). Unfortunately, only a tiny fraction of these potentially hazardous volcanoes is presently subjected to routine surveillance. Indeed, it is only since the advent of remote sensing that volcanologists have attempted merely to identify and catalogue all the world's active volcanoes (e.g. de Silva and Francis, 1991).

Few, if any, volcanoes erupt without warning. Nevertheless, the signals of impending eruption are often subtle or ambiguous, and eruption prediction, which is essentially based on empiricism and pattern recognition, is rarely attempted. While the most rigorous investigations tend to be conducted at frequently erupting volcanoes, such as Mount Etna and Kilauea, a large proportion of the most devastating eruptions has taken place at volcanoes that had been quiescent for centuries or millenia (e.g. Dvorak *et al.*, 1990). In 1956, 3000 people in Papua New Guinea were killed by an eruption of Mt Lamington, which was not even known to be a volcano (Taylor, 1983). With the UN-endorsed International Decade of Natural Disaster Reduction under way, remote sensing satellites, in consort with advances in geophysical and geochemical surveillance, are contributing increasingly towards both the operational needs of volcano monitoring and emergency response, and to a better understanding of the physical processes underlying volcanic behaviour.

Current applications of remote sensing techniques to volcanic studies

Development of modern satellite remote sensing techniques has opened up several entirely new fields of volcanic research and volcano monitoring. While much has been accomplished, existing satellite sensor systems were designed for quite different purposes, notably in the fields of agriculture and meteorology, and their volcanological results are therefore merely incidental. In future decades, sensor systems designed with specific volcanic applications in mind will offer new potential for volcano monitoring. The present period could be characterized as an inter-regnum: although the potential of present and future satellite systems is understood, the volume of data available is limited and costly. By the end of this decade, the situation may be much improved. Because aspects of these issues have already been adequately addressed elsewhere (for example, Mouginis-Mark et al., 1989; Francis, 1989), we provide here a guide to the existing literature, review some of the successes of volcanological remote sensing techniques, discuss some of their limitations and describe some future applications.

The utility of satellite remote sensing systems for volcanology stems from four essential aspects:

(1) global coverage,
(2) synoptic perspective,
(3) extension of the perceptual range through many parts of the electromagnetic spectrum, including the ultraviolet, visible, infrared and microwave, and
(4) continuity of data acquisition.

Examples of all these attributes are illustrated below.

Mapping

When the first LANDSAT was launched in 1972, volcanologists had access for the first time to image data for all the world's volcanoes, not merely those in an individual region or national territory. Inevitably, the value of LANDSAT data was greatest in regions where the least was known already, and where conventional mapping had made the least impression—a remarkably large part of the world. On geological surveys around the world, LANDSAT MSS images were seized on as reliable mapping tools. Because of their limited spectral capability, MSS data were used primarily for topographic and morphological interpretations—images were used essentially as large-scale air photographs.

Even this unsophisticated application led to some important discoveries. In the course of a volcanological investigation of LANDSAT MSS images of the central Andes, many large, previously unknown, volcanic structures were found (Francis and Baker, 1978; Baker, 1981). Some of these led to several years of follow-up field investigations, for example the 2.2 my old Cerro Galan caldera in northwest Argentina (Figures 4.1a, 4.1b), arguably the best exposed example of a large resurgent caldera on Earth (Francis *et al.*, 1983). Other important structures still remain to be explored on the ground. Although the LANDSAT MSS transformed regional volcanological studies, it remained of limited use for more detailed investigations by its 80 m pixel size and restricted spectral range (four bands in the visible and near infrared). When the Thematic Mapper (TM) became operational on LANDSATs 4 and 5, more subtle discrimination of geomorphic features and volcanogenic products became possible, enabling relative dating of lavas, pyroclastic and debris flows, and airfall deposits. For example, the presence of an unusual

Figure 4.1(a)   LANDSAT MSS image (Band 7) acquired 15 March 1976 of the Cerro Galan caldera, northwest Argentina (26°S, 67°W). Caldera is 35 × 20 km; snow covered resurgent centre is 6000 m high. This image was instrumental in the discovery and mapping of the previously unknown caldera.
(NASA ERTS E 2418-13393-7)

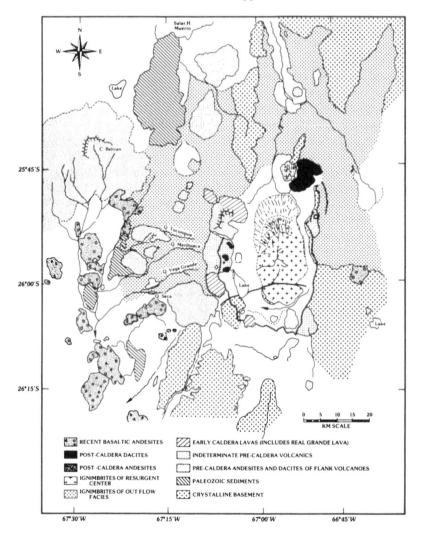

Figure 4.1(b)   Outline geological map constructed from Figure 4.1(a) and ground truth obtained by field expeditions. (Francis et al., 1978, 1983).

deposit south of Tata Sabaya volcano, Bolivia, had been noticed on MSS images. When TM data were available (Figure 4.2), the improved spatial resolution revealed unambiguously the hummocky topography definitive of a large volcanic debris avalanche deposit (Francis and Wells, 1988). Studies of these sorts, dependent on morphological analyses, are likely to prove increasingly useful in the construction of hazard zonation maps of potentially active volcanoes, whose utility is based on characterizing the volcanoes' past eruptive histories.

Figure 4.2 LANDSAT TM image of the Tata Sabaya volcano, Bolivia, and its debris avalanche deposit, extending 20 km into the Salar de Coipasa. Originally noted on MSS images, the improved spatial resolution of the TM revealed unambiguously the nature of the deposit. Follow-up field studies suggest that the avalanche was emplaced about 10 000 years ago. (LANDSAT TM FCC, bands 7,4,2).

Figure 4.3 LANDSAT TM image of 6000 m high San Pedro volcano, north Chile, illustrating a range of volcanological phenomena identifiable on 30 m pixel images. Young lavas (bottom right and centre); pyroclastic flow deposits (top right); a young scoria cone and lava flow (left centre) and an older debris avalanche deposit (top left) are all evident. Image is 20 km across. (LANDSAT TM FCC, bands 7,4,2).

Figure 4.4   *LANDSAT TM image of the 24 km³ Chao lava, north Chile, showing surface texture and massive ogive ridges. Synoptic image reveals Chao vent lying along same NW–SE trending structural lineament as older adjacent volcanoes (snow covered). Trend is oblique to the regional trend of Andean cordillera. Image is 45 km across.*
(LANDSAT TM FCC, bands 7,4,2)

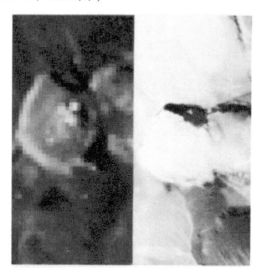

Figure 4.6   *(Left) Example of thermal emission detected in TM bands 5 and 7 from a volcanic source: Láscar volcano, north Chile. The image shows that thermal anomalies only a few pixels across can be unambiguously detected, while the structure of the anomaly provides evidence of the nature of the volcanic manifestation. In this case, radiant anomalies are believed to be due to rocks heated by fumaroles at magmatic temperatures (~ 1000°C). Crater is about 800 m across.*
(Right) Non-contemporaneous vertical air photograph of Láscar, at same scale. Although thermal sources were probably present, the air photograph provides no evidence for them, emphasizing the utility of satellite SWIR studies.

In order to identify all the potentially active volcanoes in the central Andes, 29 TM scenes were obtained, covering the entire volcanic province of the region (de Silva and Francis, 1991). Fieldwork supported by conventional air photography was carried out in a number of selected areas to provide ground truth for the TM data. Supplementary data sources which proved useful include the experimental Modular Optoelectronic Multispectral Scanner (MOMS-01) flown on Space Shuttle missions STS-7 (June 1983) and STS-11 (February 1984). The most useful attribute of the MOMS data proved to be the serendipitous very low angle, westerly illumination (Rothery and Francis, 1987). SPOT images of two volcanoes (Socompa and Llullaillaco, both in Chile) were also acquired. The 20 m multispectral SPOT data proved to have little advantage over the TM data, but the 10 m panchromatic stereo SPOT images represented a powerful additional means of making morphological and textural studies.

Inevitably, the extent of remote sensing data for this remote region led to many useful insights. For example, the synoptic view of the TM data enabled large volcanic complexes to be studied in single image quadrants, facilitating analyses of structural relationships between volcanoes and regional tectonic elements (e.g. Allmendinger *et al.*, 1989). Most importantly, however, the study led to the identification of thirty potentially active major volcanoes, including many over 6000 m high, and numerous minor structures, many of them previously entirely unknown (Francis and de Silva, 1989; de Silva and Francis, 1990). Improved understanding of the structures and histories of previously known volcanic structures was also obtained (Figures 4.3 and 4.4). Hand-held Shuttle Hasselblad photographs were also used in a subsidiary detailed study of large volcanic debris avalanches in the central Andes. Fifteen previously unknown avalanche deposits were discovered using a combination of TM images and Shuttle photography (Francis *et al.*, 1985; Francis and Wells, 1988).

Several other groups have carried out similar studies. For example, Munro and Mouginis-Mark (1990) combined SPOT-1 HRV and large format camera images in a study of Fernandina volcano in the Galapagos Islands, and Bonneville *et al.* (1989) have mapped the principal volcanic units of Piton de la Fournaise, Réunion Island. Other groups are integrating digital elevation models (DEMs) with satellite data within the GIS environment (Young and Wadge, 1990). With the extra dimension of surface topography, questions concerning matters such as slope stability and the likely paths of lava or other flows can be addressed. Automated computer techniques are even available for obtaining DEMs directly from stereoscopic satellite image pairs such as those from SPOT.

Detecting and measuring volcanic thermal emissions

For over 25 years, airborne and ground-based thermal infrared (TIR) surveys have been carried out at numerous active volcanoes and geothermal sites for the purpose of detection, and monitoring of thermal anomalies (e.g. Fischer *et al.*, 1964; Moxham, 1971; Kieffer *et al.*, 1980; Bianchi *et al.*, 1990). There are two levels of investigation of such data: the first is concerned simply with detection of new thermal features or changes in existing ones; the second asks whether the data have a more quantitative value for modelling radiative parameters which would benefit both theoretical studies of physical processes in volcanic activity, and have enhanced potential for hazard assessment. However, quantitative measurements from these kinds of data are not straightforward because of difficulties such as calibration of the instruments and correction for effects of solar heating and reflection. Airborne campaigns are costly to mount and pose many logistical problems. They have not provided routine sources of infrared data for volcano monitoring nor are they likely to do so in the future.

Orbital remote sensing from unmanned spacecraft does, however, offer a means of regular observation of subaerial volcanoes world-wide (Francis, 1979), and the potential of on-board infrared sensors to detect thermal manifestations was demonstrated as long ago as 1966 when the High Resolution Infrared Radiometer (3.45–4.07 μm) on board the Nimbus II meteorological satellite detected the thermal signature of lava flows erupted at Surtsey, Iceland (Williams and Friedman, 1970). More recently, Bonneville *et al.* (1985) and Bonneville and Kerr (1987) have used TIR satellite data (recorded by the Advanced Very High Resolution Radiometer, AVHRR, on board the NOAA-series platforms) to map thermal anomalies on Mount Etna, Italy. If such measurements can be calibrated, radiant temperatures may be estimated for the surface viewed whether it be ground, sea, cloud, ash plume, or other, using the underlying principle of Planck's distribution law which expresses radiated power output as a function of temperature and wavelength.

Planck's law also underpins a two-waveband technique for estimating temperatures and sizes of sub-pixel thermal features (Dozier, 1981; Matson and Dozier, 1981). Following discovery of a pronounced thermal anomaly in the short-wavelength infrared (SWIR) bands of TM images of the north Chilean volcano, Láscar (Francis and Rothery, 1987), Rothery *et al.* (1988) adapted this technique to estimate radiative temperatures at several volcanoes, including the remote Erta 'Ale (Ethiopia) and Erebus (Antarctica). Observations in the SWIR part of the spectrum are distinct from those at longer wavelengths because of its sensitivity to very high temperature (magmatic) phenomena. As Figure 4.5 demonstrates, it is possible to detect in the two TM bands 5 and 7, ground at temperatures around 1000°C which occupies just one-ten thousandth of the nominal 30 × 30 m pixel area (Rothery *et al.*, 1988). These early studies were extended to model changes in radiated power output at Láscar volcano (Francis *et al.*, 1989, Glaze *et al.*, 1989 a,b), and estimate heat losses from lava flows at Mount Etna (Pieri *et al.*, 1990) and the Chilean volcano, Lonquimay (Oppenheimer, 1991).

Current research is directed towards understanding how the spatial and

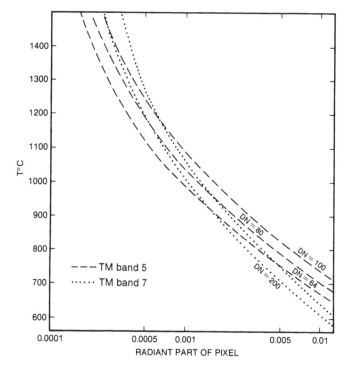

Figure 4.5 Graphical examples of the two wave-length technique of Matson and Dozier (1981), applied to TM data for determination of temperature and fractional part of radiant pixels occupied by volcanic thermal anomalies. Curves for different DNs in TM bands 5 and 7 are shown; intersections of curves define temperature and radiant part of pixel for each case.

spectral characteristics of such anomalies can best be interpreted in terms of volcanic phenomena, for example, distinguishing between fumaroles at magmatic temperatures, and surface extrusions of lava (Figure 4.6). For the present, detection of magmatic thermal anomalies on remote volcanoes provides a powerful new means of identifying new events and has clear implications in hazard assessment.

Observations of explosive volcanic eruptions

Volcanic ash clouds

Because they are so much more extensive than volcanoes themselves, volcanic eruption clouds are easily detected on meteorological satellite scenes. Even on the whole hemisphere images provided by geostationary satellites such as the GOES (Geostationary Operational Environmental Satellite) series, it is possible to recognize eruption clouds even though pixels are about 4 km across. Many studies of the downwind transport and dispersal of eruption clouds have been made (e.g. Kienle and Shaw, 1979; Sawada, 1983; Robock and Matson, 1983; Malingreau and Kaswanda, 1986). More recently, efforts have been directed to identify spectral differences in the visible and infrared ranges which enable the discrimination of smaller plumes from meteorological clouds (e.g. Prata, 1989). Eruption plumes have also often been imaged by higher resolution sensors such as the TM, notably in the case of the eruption of Augustine, Alaska on 27 March 1986. However, the lengthy repeat acquisition cycles dictate that these are single 'snapshots', of limited value for sustained monitoring of the course of an eruption. Data from geostationary weather satellites and polar orbiters such as the NOAA-series have proved most valuable for continuous monitoring and rapid response, notwithstanding their coarser spatial resolution.

A minor eruption of Láscar volcano, north Chile, exemplifies the level of detail that has been achieved with existing technology (Glaze *et al.*, 1989b). Images from the GOES-West were used. This system images most of the western hemisphere every 30 minutes in two spectral bands (visible, 0.55–0.75 μm and thermal infrared, 10.5–12.5 μm). Pixels in the thermal band have dimensions of 1.16×1.16 km near Láscar, while thermal infrared pixels are 4×8 km. The GOES images showed that the eruption occurred at about 10.39 UT on 16 September 1986 and lasted less than five minutes. An ash cloud climbed rapidly to a height of about 15 km above the volcano, and began to drift rapidly downwind, southeastwards. Average velocities at the head of the plume, as determined from its position on successive images, reached 181 km h^{-1}. Shadow studies indicated that the ash cloud was 3–4 km thick and that its elevation varied from a maximum of 15 km to about 10 km. By 12.12 UT, the plume began to pass overhead the city of Salta, Argentina, 285 km from the volcano, where ash-fall was detected. During the three and a half hours following the eruption, the plume travelled a minimum of 400 km and expanded to cover an area greater than 100 000 km^2.

These observations are noteworthy in three respects: first, remote sensing data are the only sources of information on the eruption, which came to light only because of the reported ash-fall on Salta. Second, they provide an invaluable means of testing volcanological models for predicting the fall-out of ash from eruptions. Third, from the standpoint of volcanic hazards, they highlight the risks of volcanic ash clouds to aircraft (Kienle *et al.*, 1990). Recent incidents in which large passenger aircraft have encountered serious difficulties in volcanic ash plumes took place during the eruptions of Galunggung (Indonesia, 1982), Redoubt (Alaska, 1990) and Pinatubo (Philippines, 1991). In the case of the Láscar eruption—a minor one—ash had drifted into air corridors hundreds of kilometres from the volcano less than three hours after the eruption.

Most recently, infrared images from the NOAA-10 satellite were used on 15 June 1991 to reveal the extent of the massive ash cloud from the Pinatubo (Philippines) eruption. At about 18.30 the eruption column rose to a height of 35 to 40 km immediately over the volcano and extended almost 1000 km downwind. Numerous aircraft reported damage and loss of engine power from ash encounters, but no accidents ensued (Smithsonian Institution, 1991).

It is self-evident that rapid and reliable means of alerting aircraft to the track of ash clouds are urgently required. Implementation of such procedures is under way between the National Oceanographic and Oceanic Administration and the Federal Aviation Authority in the USA.

Volcanic gas clouds

In addition to silicate ash, major explosive eruptions inject large amounts of gases into the atmosphere. Even quiescent volcanoes are notable for their gas emissions and large volume effusive eruptions have also been implicated in substantial release of volatile gases to high altitudes (Self *et al.*, 1981). In particular, sulphur dioxide released from volcanoes influences the physicochemical nature of the atmosphere over long periods by formation of sulphuric acid aerosols. Sulphur dioxide absorbs certain wavelengths of ultraviolet radiation between 0.30 and 0.33 μm, and it this property which has enabled the Total Ozone Mapping Spectrometer (TOMS) carried by the Nimbus 7 satellite to detect and quantify volcanogenic releases. This is another inspired use of a sensor originally designed for a quite different purpose.

Major eruptions from El Chichón (a previously little-known volcano) in March and April 1982 propelled eruption columns more than 20 km high into the atmosphere. Many ground-based and satellite observations (including NOAA-6 and -7, GOES-East and -West, and METEOSAT) tracked the eruption plume, which ultimately spread right around the world within a narrow latitudinal zone (Robock and Matson, 1983). On the basis of spectrometer data and the extent of the cloud, Krueger (1983) estimated that the volcano injected 3.3 million tonnes (Mt) of gaseous sulphur dioxide into the stratosphere, and that within three months of the eruption all of this gas had been converted into sulphuric acid aerosols. TOMS data for the 1985 eruption of Nevado del Ruiz suggested a lesser release of sulphur dioxide, 0.66 Mt (Krueger *et al.*, 1990), but the 1991 episode at Pinatubo is believed to have exceeded the El Chichón eruption in sulphur dioxide release by a factor of two (Smithsonian Institution, 1991).

These observations are important because they have a bearing on the climatic effects of large volcanic eruptions—El Chichón may have caused a 0.2°C drop in temperature in June 1982 and has been implicated in the subsequent El Niño event (Rampino and Self, 1984). Sulphur dioxide plumes from many other eruptions have been detected by the TOMS sensor. What is perhaps more surprising, however, is that many other eruptions have *not* been detected. There may be many reasons for this, some of which, such as variations in the sulphur content of the original magma, are understood. Other factors, such as the mechanisms of injection into the stratosphere by different kinds of eruption, and residence time of gases and aerosols in the stratosphere, are less well understood and are the subjects of on-going research.

Limitations of existing techniques

Although much has been achieved, several simple but profound limitations restrict the routine application of satellite remote sensing to volcanology: access to data, clouds and scarcity of ground truth data.

Access to data

There are both physical and financial restrictions to high-resolution data from the LANDSAT and SPOT systems. LANDSAT 5, for example, nominally provides coverage of an individual volcano only once every 16 days. (Coverage is more frequent at high latitudes where ground tracks converge.) Much can happen on a volcano within that time. While SPOT has the potential for more frequent cover (five-day repeat cycle), it lacks the infrared capability of the LANDSAT TM. Furthermore, costs of acquiring data from commercial remote sensing institutions are prohibitive for most academic purposes. To our knowledge, the longest time-series coverage of high-resolution images covers Láscar, which has been imaged 14 times between 1984 and 1991. Even where funds are available, dissemination of TM data is often protracted, so that all such investigations are retrospective.

Clouds

The gravity of this elementary problem is best illustrated by an example: to date, it has not been possible to obtain useful cloud-free TM images of Kilauea volcano in Hawaii. Many other volcanoes in humid tropical and temperate locations are similarly affected. Unfortunately, the implication of this is that visible and near infrared remote sensing techniques can never be relied on to provide continuous monitoring of volcanoes. At best, they can only provide valuable complements to other techniques, notably satellite-borne SARs, and conventional ground techniques.

Ground truth

As in every other aspect of remote sensing, volcanological interpretations are only as good as the data that constrain them. While satellite remote sensing can provide images for all of the world's subaerial volcanoes, the background knowledge of these volcanoes and their characteristic behaviour varies tremendously. Manifestations of activity observed on a well-known volcano in a temperate environment may be confidently interpreted, but apparently similar manifestations on an unknown volcano in a different environment may be erroneously interpreted. Even seasonal differences, such as variations in snow cover, can enormously complicate interpretations. Where assumptions are made in order to make measurements from 'raw' satellite data, again, the results are only as dependable as the applied models are realistic. For example, to make estimates of sub-pixel scale temperatures by two-wavelength techniques requires the assumption that very simple two component thermal distributions prevail over the surfaces imaged. Only limited ground truth is available to lend support to the applicability of such approximations (e.g. Flynn *et al.*, 1989; Oppenheimer and Rothery, 1991).

A similar, but even more serious problem concerns volcanic eruption plumes. Because of their importance in global atmospheric studies, these are likely to be observed much more closely in the coming decade. Indeed, the improved spectral potential of new sensor system offers remarkable new opportunities for determining the compositions of plumes and the proportions of constituent gases, particulates and aerosols. At present, however, little is known of the spectral properties of ash clouds, nor is there much useful 'cloud truth' data against which to check remote sensing measurements.

Future potential of satellite remote sensing systems in volcanology

Numerous future applications of remote sensing techniques to volcanology have been explored, notably within the framework of the EOS (Earth Observing System) Interdisciplinary Volcanology project (Mouginis-Mark *et al.*, 1991; Mouginis-Mark and Francis, in press). Given the number and diversity of satellite systems to be launched in the coming decade (ERS-1, JERS, SIR-C, EOS-A, etc.), it is certain that vast amounts of new data will become available. Indeed, the sheer volume of the data stream that will ultimately be available from EOS will inevitably present management and dissemination problems. In terms of pure research, this richness of data can only be desirable and will certainly spawn numerous new lines of enquiry. However, budgetary uncertainties mean that it is difficult to predict when these new lines of investigation will become practical.

Monitoring of volcanoes to contribute towards hazards assessment requires a completely different approach from strictly academic lines of enquiry. For the reasons outlined above, visible and near-infrared are unlikely to be reliable in this role, except in the domain of ash plume tracking. Imaging radar systems offer great promise since they are unaffected by cloud cover and can obtain images by day or night. They can provide data of similar spatial detail as TM images, and could reveal, for example, the eruption of a new lava flow or fresh ash. More importantly, radar interferometric techniques are capable of detecting very subtle topographic changes—a few centimetres or less—offering the potential to detect ground deformation on and around volcanoes, of the order of magnitude known to be associated with impending volcanic eruptions.

References

Allmendinger, R.W., Stretcher, M., Eremchuk, J.E. and Francis, P.W., 1989, Neotectonic deformation of the southern Puna Plateau, NW Argentina, *Journal of South American Earth Sciences*, **2**, 111-30.

Baker, M.C.W., 1981, The nature and distribution of Upper Cenozoic ignimbrite centres in the Central Andes, *Journal of Volcanology and Geothermal Research*, **11**, 293-315.

Bianchi, R., Casacchia, R., Coradini, A., Duncan, A.M., Guest, J.E., Kahle, A., Lanciano, P., Pieri, D.C., Poscolieri, M., 1990, Remote sensing of Italian volcanoes, *EOS, Transactions of the American Geophysical Union*, **71**(46), 1789-91.

Bonneville, A. and Kerr, Y., 1987, A thermal forerunner of the 28th March 1983 Mt Etna eruption from Satellite Thermal Infrared Data, *Journal of Geodynamics*, **7**, 1-31.

Bonneville, A., Vasseur, G. and Kerr, Y., 1985, Satellite Thermal Infrared Observations of Mt Etna after the 17th March 1981 Eruption, *Journal of Volcanology and Geothermal Research*, **24**, 293-313.

Bonneville, A., Lanquiette, A.M., Pejoux, R. and Bayon, C., 1989, Reconnaissance des Principales Units Geologiques de Piton de la Fournaise, a partir de SPOT-1. *Bulletin Societé Geologique, France*, **8**, 1101-10.

de Silva, S.L and Francis, P.W., 1990, Potentially active volcanoes of Southern Peru, *Bulletin of Volcanology*, **52**, 286-301.

de Silva, S.L. and Francis, P.W., 1991, *Volcanoes of the Central Andes*, Berlin: Springer Verlag, 216pp.

Dozier, J., 1981, A method for satellite identification of surface temperature fields of sub-pixel resolution, *Remote Sensing of Environment*, **11**, 221-9.

Dvorak, J., Matahelumual, J., Okamura, A.T., Said, H., Casadevall, T.J. and Mulyadi, D., 1990, Recent uplift and hydrothermal activity at Tangkuban Parahu Volcano, West Java, Indonesia, *Bulletin of Volcanology*, **53**, 20-28.

Fischer, W.A., Moxham, R.M., Polcyn, F. and Landis, G.H., 1964, Infrared surveys of Hawaiian volcanoes, *Science*, **146**, 733-42.

Flynn, L.P., Mouginis-Mark, P.J. and Gradie, J.C., 1989, Radiative temperature measurements at Kilauea Volcano, Hawaii. Abstract, IAVCEI General Assembly, Santa Fe, 25 June–1 July 1989, *Bulletin of the New Mexico Bureau of Mines and Mineral Resources*, **131**, 94.

Francis, P.W., 1979, Infrared techniques for volcano monitoring and prediction—a review, *Journal of the Geological Society of London*, **136**, 355-60.

Francis, P.W., 1989, Remote sensing of volcanoes, *Advances in Space Research*, **9**, 89-92.

Francis, P.W., and Baker, M.C.W., 1978, Sources of two large volume ignimbrites in the Central Andes: some LANDSAT evidence, *Journal of Volcanology and Geothermal Research*, **4**, 81-7.

Francis, P.W. and de Silva, S., 1989, Application of the LANDSAT Thematic Mapper to the identification of potentially active volcanoes in the Central Andes, *Remote Sensing of the Environment*, **28**, 245-55.

Francis, P.W. and Rothery, D.A., 1987, Using the LANDSAT Thematic Mapper to detect and monitor active volcanoes: an example from the Láscar Volcano, Northern Chile, *Geology*, **15**, 614-17.

Francis, P.W. and Wells, G.L., 1988, LANDSAT Thematic Mapper Observations of large volcanic debris avalanche deposits in the Central Andes, *Bulletin of Volcanology*, **50**, 258-78.

Francis, P.W., Hammill, M., Kretzschmar, G.A. and Thorpe, R.S., 1978, The Cerro Galan Caldera, Northwest Argentina, *Nature*, **274**, 749-51.

Francis, P.W., Kretschmar, G., O'Callaghan, L., Thorpe, R.S., Sparks, R.S.J., Page, R., de Barrio, R.E., Guillou, G. and Gonzalez, O., 1983, The Cerro Galan Ignimbrite, *Nature*, **310**, 51-53.

Francis, P.W., Gardeweg, M., O'Callaghan, L.J., Ramirez, C.F. and Rothery, D.A., 1985, Catastrophic debris avalanche deposit of Socompa Volcano, North Chile, *Geology*, **13**, 600-3.

Francis, P.W., Glaze L.S. and Rothery, D.A., 1989, Láscar Volcano set to erupt, *Nature*, **339**, 434.

Glaze, L.S., Francis, P.W. and Rothery, D.A., 1989a, Measuring thermal budgets of active volcanoes by satellite remote sensing, *Nature*, **338**, 144-6.

Glaze, L.S. Francis, P.W., Self, S. and Rothery, D.A., 1989b, The Láscar September 16 1986 eruption: satellite investigations, *Bulletin of Volcanology*, **51**, 149-60.

Kieffer, H.H., Frank D. and Friedman, J.D., 1980, Thermal Infrared Surveys at Mount St. Helens—Observations prior to the eruption of May 18, in Lipman, P.W. and Mullineaux, D.R. (eds), The 1980 Eruptions of Mount St. Helens, Washington, *US Geological Survey Professional Paper*, **1250**, 257-77.

Kienle, J. and Shaw, G.E., 1979, Plume dynamics, thermal energy and long-distance transport of vulcanian eruption clouds from Augustine Volcano, Alaska, *Journal of Volcanology and Geothermal Research*, **27**, 179-94.

Kienle, J., Dean, K.G., Garbeil, H. and Rose, W.I., 1990, Satellite surveillance of volcanic ash plumes, applications to aircraft safety, *EOS, Transactions of the American Geophysical Union*, **71**, 266.

Krueger A.J., 1983, Sighting of El Chichón sulfur dioxide clouds with the Nimbus 7 Total Ozone Mapping Spectrometer, *Science*, **220**, 1377-9.

Krueger A.J., Walter L.S., Schnetzler C.C. and Doiron S.D., 1990, TOMS measurement of the sulfur dioxide emitted during the 1985 Nevado del Ruiz eruptions, *Journal of Volcanology and Geothermal Research*, **41**, 7-15.

Malingreau, J.P. and Kaswanda, 1986, Monitoring volcanic eruptions in Indonesia using weather satellite data: the Colo Eruption of July 28, 1983, *Journal of Volcanology and Geothermal Research*, **27**, 179-94.

Matson, M. and Dozier, J., 1981, Identification of subresolution high temperature sources using a thermal IR sensor, *Photogrammetric Engineering and Remote Sensing*, **47**, 1311-8.

Mouginis-Mark, P.J. and Francis, P.W. submitted, Satellite observations of active volcanoes: prospects for the 1990s, *Episodes*.

Mouginis-Mark, P., Pieri, D.C., Francis, P.W., Self, S. and Wood, C.A., 1989, Remote sensing of volcanoes and volcanic terrains, *EOS, Transactions of the American Geophysical Union*, **70**, 1567-75.

Mouginis-Mark, P., Rowland, S., Francis, P.W., Friedman, T., Gradie, J., Self, S. Wilson, L., Crisp, J., Glaze, L., Jones, K., Kahle, A., Pierie, D., Zebker, H., Kreuger, A., Walter, L., Wood, C., Rose, W., Adams, J. and Wolff, R., 1991, Analysis of active volcanoes from the Earth Observing System, *Remote Sensing of Environment*, **36**, 1-12.

Moxham R.M., 1971, *Thermal Surveillance of Volcanic Activity*, Paris: UNESCO, 103-24.

Munro, D.C. and Mouginis-Mark P.J., 1990, Eruptive patterns and structure of Isla Fernandina, Galapagos Islands, from SPOT-1 HRV and large format camera images, *International Journal of Remote Sensing*, **11**, 1501-9.

Oppenheimer C., 1991, Lava flow cooling estimated from LANDSAT Thematic Mapper data: the Lonquimay Eruption (Chile, 1989), *Journal of Geophysical Research*, **96**, 21865-21878.

Oppenheimer, C. and Rothery, D.A., 1991, Infrared monitoring of volcanoes by satellite, *Journal of the Geological Society of London*, **148**, 563-9.

Pieri, D.C., Glaze, L.S. and Abrams, M.J., 1990, Thermal radiance observations of an active lava flow during the June 1984 Eruption of Mount Etna, *Geology*, **18**, 1018-22.

Prata A.J., 1989, Observations of volcanic ash clouds in the 10-12 μm window using AVHRR/2 data, *International Journal of Remote Sensing*, **10**, 751-61.

Rampino, M.R. and Self, S., 1984, The atmospheric effects of El Chichón, *Scientific American*, **250**, 48-57.

Robock, A. and Matson, M., 1983, Circumglobal transport of the El Chichón volcanic dust cloud, *Science*, **21**, 195-7.

Rothery, D.A. and Francis, P.W., 1987, Synergistic use of MOMS and TM data, *International Journal of Remote Sensing*, **8**, 501-8.

Rothery, D.A., Francis, P.W. and Wood, C.A., 1988, Volcano monitoring using short wavelength infrared data from satellites, *Journal of Geophysical Research*, **93**, 7993-8008.

Sawada, Y., 1983, Analysis of eruption clouds by the 1981 eruptions of Alaid and Pagan volcanoes with GOES images, *Papers in Meteorology and Geophysics*, **34**, 307-24.

Self, S., Rampino, M.R. and Barbera, J.J., 1981, The possible effects of large 19th and 20th Century volcanic eruptions on zonal and hemispheric surface temperatures, *Journal of Volcanology and Geothermal Research*, **11**, 41-60.

Smithsonian Institution, 1991, Pinatubo, *Bulletin of the Global Volcanism Network*, **16**, 2-8.

Taylor, G.A.M., 1983, The 1951 eruption of Mt. Lamington, Papua, *Bulletin of the Bureau of Mineral Resources, Geology and Geophysics*, Canberra, Australia, **38**, 129pp.

Tilling, R.I., 1989, Volcanic hazards and their mitigation: progress and problems, *Reviews of Geophysics*, **27**, 237-69

Williams, R.S., Jr. and Friedman J.D., 1970, Satellite observation of effusive volcanism, *Journal of the British Interplanetary Society*, **23**, 441-50.

Young, P. and Wadge, G., 1990, Flowfront: simulation of a lava flow, *Computers and Geosciences*, **16**, 1171-91.

Chapter 5
Soil/vegetation characteristics at microwave wavelengths

A.M. SHUTKO

Institute of Radio Engineering and Electronics,
Academy of Sciences of the USSR, Moscow

Introduction

Microwave radiometry is an effective method for Earth surface monitoring developed at the Institute of Radio Engineering and Electronics of the Academy of Sciences of the USSR. This paper briefly discusses the principles underlying the method and its feasibility for the determination of specific soil and vegetation parameters. The effectiveness of the combined use of microwave, infrared and optical data is demonstrated, and examples of the practical application of the method are presented.

Physical background

The passive microwave (MCW) or super-high-frequency (SHF) band of the electromagnetic spectrum ranges from millimetre to centimetre and decimetre wavelengths. In practice the most favourable band for land surface studies covers the wavelengths from 0.8-2 cm to 20-30 cm. Clouds and rain are the main sources of interference at shorter wavelengths and galactic and ionospheric radiation are the main sources of interference at longer wavelengths (Shutko, 1986, 1987). Within this range of wavelengths, radiation is measured in terms of the brightness temperature T_B, which is the product of emissivity and effective physical temperature, and is more or less sensitive to the moisture content, density, salinity and temperature of the soil, as well as to the type, biomass and temperature of vegetation. The sensitivity to some parameters is dependent on wavelength. The schematic diagrams in Figure 5.1 and the real data presented in Figures 5.2 to 5.8 give an impression of how the main parameters mentioned above influence radiation at the wavelengths most typically used in land surface studies, i.e. 0.8-2 cm and 20-30 cm (Armand *et al.*, 1989; Basharinov and Shutko, 1975; Chukhlantsev *et al.*, 1989; Kirdiashev *et al.*, 1979; Shutko, 1986, 1987; Shutko and Chukhlantsev, 1982).

It is seen, in particular, that the intensity of radiation decreases when there is an increase of soil moisture and mineralization (salinity) of open water and soil water. However, for dry soil the presence of salts does not result in a corresponding decrease in brightness temperature. The presence of near-surface groundwater (i.e. where the water table is shallow) up to 3 to 4 m deep first manifests itself in a decrease of T_B at decimetre wavelengths.

Vegetation does not cause significant changes in T_B for dry soil. Here one may observe either a small increase or a decrease of emissivity *æ*. For broad-leaved crops, the direction of change depends on whether or not reflection from leaves takes place. Where a very moist soil or water surface underlie the vegetation canopy, a notable increase in T_B contrast is a function of wavelength, type and biomass of vegetation cover. Figures 5.2 to 5.8 show estimates of the sensitivity of T_B to variations in the values measured by physical parameters.

Three different types of water are found in soil: tightly-bonded, loosely-

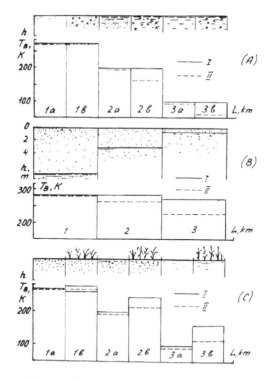

Figure 5.1   Schematic diagrams of brightness temperature (T_B) response at 0.8–2 cm (I) and 20–30 cm (II) wavelength to variations of some characteristic parameters of soil, water, and vegetation: (A) T_B changes for dry soil (1), wet soil (2) and open water (3) in the absence (a) and presence (b) of salts in media; (B) T_B changes for different levels of groundwater (shallow water table); (C) T_B changes for dry soil (1), wet soil (2) and water (3) in the absence (a) and presence (b) of vegetation.

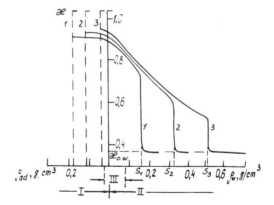

Figure 5.2 Microwave emissivity as a function of adsorbed water in soil (I), free water (II) and loosely-bonded water (III) for silty soil (1), clay soil (2) and sandy soil (3).
æ_{o.w.} *is the emissitivity of open water; S_1, S_2, S_3 are the values of free water content under saturation.*

bonded, and free water (Figure 5.2). The dielectric permittivity, ε, of adsorbed water is similar to that of particles of dry soil, whereas the dielectric permittivity of free water, ε, is about 20 times that of dry soil and adsorbed water. This explains why emissivity is most sensitive to the free water content of soil and, in addition, to the absolute (volumetric) content but not the relative content, defined as the ratio of volumetric water content and

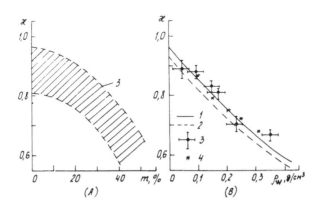

Figure 5.3 Emissivity (æ) as a function of total (adsorbed and free) water content (ρ_w) in soil, weight (m) per cent (A) and volumetric free water content (B)
(A) experimental data at wavelengths 0.8 and 3.4 cm for soil density data ranging from 0.8 to 1.7 g cm^{-3}; (B) 1, 2, theoretical data at a wavelength of 2.25 cm for soil density data 1 g cm^{-3} (1) and 1.5 g cm^{-3} (2); 3, data from field measurements; 4, laboratory measurements.

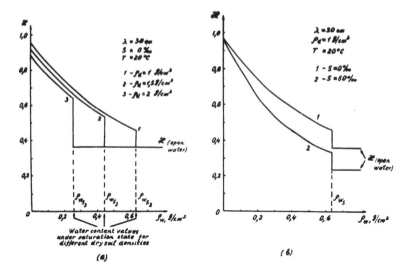

Figure 5.4　Emissivity (æ) as a function of free water content (ρ_w) under variations of soil density (a) and mineralization (salinity) of water in soil (b) at $\lambda = 30$ cm for temperature $T = 20°C$.
(A) salinity $S = 0$ ppt; 1, 2, 3, soil density data corresponding to values of 1, 1.5 and 2 g cm^{-3}, respectively; (B) soil density $\rho_d 1 = 1$ g cm^{-3}; 1, 2, salinity data corresponding to values of 0 and 60 ppt.

density of a dry soil. Thus, the boundary between adsorbed and free water content, known as wilting point (WP), indicates the boundary between two characteristic dependencies of emissivity on water content in soil, namely a flat response, practically independent of the content of adsorbed water, and a response showing a decrease of emissivity which is a function of the free

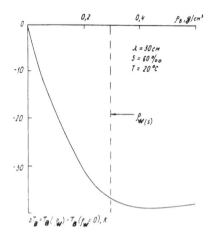

Figure 5.5　T_B contrast at 30 cm wavelength due to the presence of NaCl in water as a function of soil moisture for $T = 20°C$, $S = 60$ ppt (calculations).

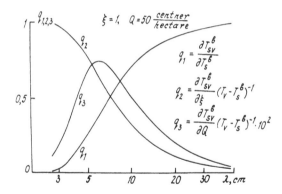

Figure 5.6 *Sensitivity of soil/canopy microwave radiation to brightness temperature (moisture content) of soil (1), density of vegetation (2) and vegetation biomass (3), for fully covered soil ($\xi = 1$) and biomass of about 50 centner per hectare (calculations)*

water content. It is clear that the type of soil is significant only in defining the boundary between these two parts of a curve. Hence, if one compares emissivity with volumetric free water content in soil there is a practically universal dependence $\text{æ}\ (\rho_W)$, independent of soil type. The slope of this curve is about -200 K/g/cm^3. On the other hand, if one compares emissivity with total water content of a soil, rather than with relative values

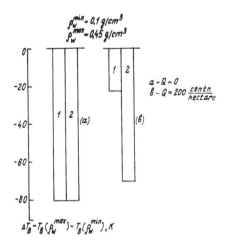

Figure 5.7 *T_B contrast due to soil moisture content increase from 0.1 to 0.45 g cm^{-3} at 2 cm (1) and 30 cm (2) wavelengths for a bare soil (a) and in the presence of vegetation (b) with biomass of about 200 centner per hectare (calculations).*

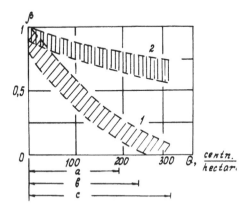

Figure 5.8 *Averaged experimental data of vegetation transmission coefficient at the wavelengths of 2.25–3.4 cm (1) and 18–20 cm (2) for narrow-leaf crops (a), cotton crop (b) and corn crop (c).*

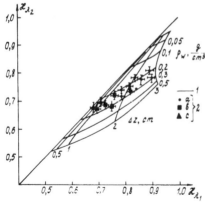
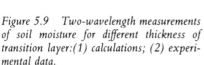

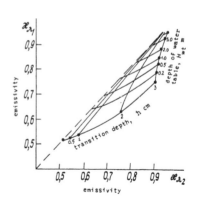

Figure 5.9   Two-wavelength measurements of soil moisture for different thickness of transition layer:(1) calculations; (2) experimental data.

Figure 5.10   Two-wavelength measurements of depth to a shallow water table for different thickness of transition layer (calculations).

such as 'weight per cent', the uncertainty of this 'radiation versus moisture content' dependence considerably exceeds the uncertainty of the dependence $\mathscr{æ}\,(\rho_W)$ (Figure 5.3).

Soil density exerts a slight influence on emissivity at all wavelengths (Figure 5.4a). The sensitivity of T_B to ρ_d is about -15 K/g/cm^3. Increasing salinity of the water in soil decreases the emissivity at decimetre wavelengths (Figure 5.4b). T_B contrast increases with increasing soil moisture until a limiting moisture value is reached (Figure 5.5). For higher values of moisture content, ΔT_B is practically independent of ρ_W and equal to that of open water. The sensitivity of T_B to changes in salinity (S) is about -0.5 K/ppt. Sensitivity to diurnal physical temperature variations at the soil surface is about 0.1 K/$^\circ$C at decimetre wavelengths and (0.3-0.4) K/$^\circ$C at centimetre wavelengths (Shutko, 1982, 1986, 1987).

Figures 5.6 to 5.8 illustrate the radiation properties of a soil/canopy system. It is seen that the screening effect of vegetation decreases at longer wavelengths. At decimetre wavelengths, narrow-leaf crops and even broad-leaf crops of biomass less than 2 kg m^{-2} are practically transparent to radiation from the soil. Sensitivity to biomass is a function of biomass.

General approach to the determination of parameters

The relationships between radiation at microwave wavelengths and the soil/vegetation complex, described in the previous section, serve as a reasonable basis for the estimation of some of parameters, in particular soil moisture (Figures 5.9 and 5.12–5.15), depth to a shallow water table (Figures 5.10 and 5.16), salinity of soil (Figure 5.11), biomass of vegetation for some different crops such as, for example, rice and cane. The methods are based on a minimum of two-wavelength measurements of the state of the observed object and use of some kind of *a priori* information about the object of interest, for example, wilting point and field capacity (FC) of the soil at some depths, type of crops and expert estimates of its biomass, etc. This kind of two-dimensional approach allows one to determine soil moisture near the surface, thickness of the transition layer (the transition being from a drier top to a wetter lower layer) (Figure 5.9) and, using measurements of WP and FC, to obtain soil moisture estimates for the top one metre layer (Figure 5.13) with an accuracy of about 0.05–0.07 g cm^{-3} (Mkrtchjan *et al.*, 1988; Reutov and Shutko, 1986, 1990; Shutko, 1985, 1986, 1987).

When considering soil moisture estimates, it is necessary to bear in mind that radiometric measurements, deriving from the explanation given above, provide the necessary data to determine absolute (volumetric) free water content in soil. The accuracy of total and relative, weight percent, free water content estimates is much lower than the accuracy for absolute free water content (Figure 5.12). *A priori* knowledge of vegetation biomass can be used to improve the accuracy of soil moisture content determination. In the absence of this kind of information, error in soil moisture estimates may exceed 0.1 g cm^{-3} (Figures 5.14 and 5.15). The accuracy of estimates of the depth to the water table is about 0.3–0.5 m in the range zero to 3–4 m in arid zones, and 0 to 1.5–2 m in moderately moist European areas (Figure 5.16).

One of the more useful sources of information about the state of environmental objects is data collected in the infrared, near-infrared and optical bands. An example of the application of multichannel measurements

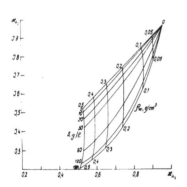

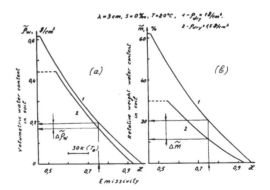

Figure 5.11 *Two-wavelength measurements of salinity of soil for different values of soil moisture (calculations).*

Figure 5.12 *Demonstration showing how the uncertainty in a priori knowledge of soil density brings one to much a smaller error in volumetric free water content estimates (a) than in relative weight per cent estimates (b). (a) and (b): 1 and 2 are different dry soil densities with values corresponding to 1 g cm^{-3} and 1.5 g cm^{-3}, respectively.*

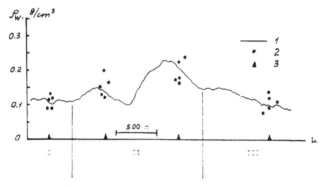

Figure 5.13   *Example of a comparison between retrieved (1) and ground truth data (2) of soil moisture content at ground stations (3) for different types of soil (I, II, III). (Bulgaria, 1987)*

in different bands, including dual-channel microwaves, is a classification of agricultural fields (Figure 5.17). Another example is dual-channel measurements used to more accurately determine soil moisture under crop cover and to obtain estimates of crop biomass (Figure 5.18).

Examples of applications

Instrumentation and associated software, designed at the Institute of Radio Engineering and Electronics of the Academy of Sciences of the USSR and its

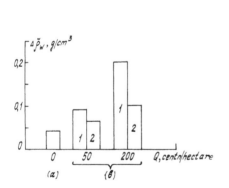

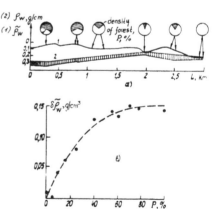

Figure 5.14   *Generalized data showing accuracy of soil moisture retrievals for bare soil (a) and soil with vegetation (b) in the absence of a priori knowledge of vegetation biomass (1) and when biomass is known to within an accuracy of 30 per cent (2).*

Figure 5.15   *Example of retrieved (1) and ground truth data (2) for soil moisture under forest canopies of different density in the absence of a priori knowledge of vegetation biomass (a) and generalized soil moisture accuracy data as a function of forest density (b).*

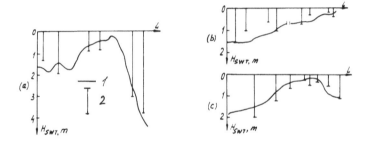

Figure 5.16 Examples of comparisons of retrieved (1) and ground truth data (2) of a depth to a shallow water table for different soil/climatic zones: (a) arid zone; (b), (c) Ukraine and Estonia, respectively.

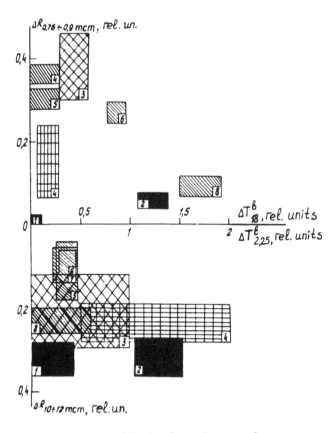

Figure 5.17 Example of agricultural fields classification by means of measurements at microwave (2.25 and 18 cm), infrared (10–12 µm), and near-infrared (0.76–0.9 µm) bands for dry soil (1), wet soil (2), grass (3), sugar beet (4) and winter wheats of different biomass (5–8).

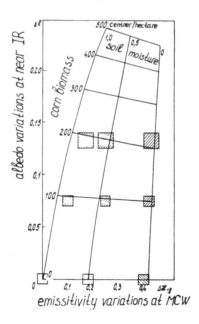

Figure 5.18 Corn biomass and soil moisture determination by means of measurements at microwaves and in near-infrared band (soil moisture is given in relative units).

Special Research Bureau, have been tested for more than 15 years in different soil-climatic zones in the USSR, from arid to humid, over an area of about 7 million hectares. The instruments are portable, stable and sensitive. Experiments have been conducted in the laboratory, in the field, from mobile platforms, and from aircraft and satellites (Shutko, 1982, 1985, 1986, 1987,

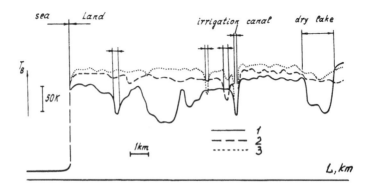

Figure 5.19 Example of spatio-temporal T_B variations as initial data for soil moisture observations during the drying process: (1) data taken just after heavy rainfall; (2) two days later; (3) seven days later.

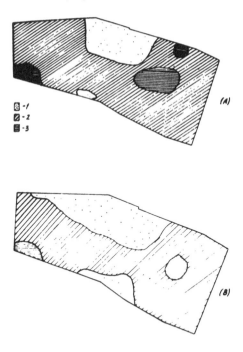

Figure 5.20   *Example of retrieved soil moisture data over the same field with an interval of one day between (B) and (A) (Ukraine).*
Soil moisture data: (1) 0.2-0.24 g cm^{-3}; (2) 0.24-0.28 g cm^{-3}; (3) 0.28-0.3 g cm^{-3}.

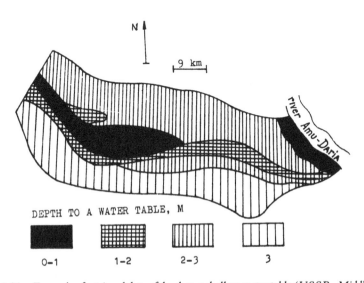

Figure 5.21   *Example of retrieved data of depth to a shallow water table (USSR, Middle Asia).*

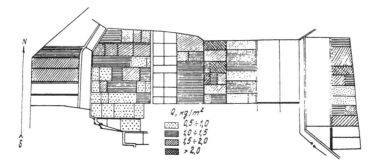

Figure 5.22   Example of retrieved data of biomass of rice crop (USSR, North Caucasus).

1990). Some operational aircraft services have been organized in different regions of the Soviet Union for agricultural, land reclamation, hydrological and agro-meteorological purposes. Hardware and software have been used on both scientific and commercial bases in Bulgaria, Hungary, Poland, Germany, Cuba, Vietnam and the USA (Shutko, 1985, 1986, 1987, 1990).

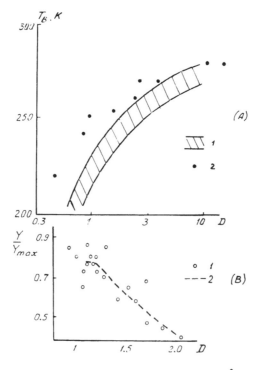

Figure 5.23   Brightness temperature (T_B) and dryness index (D) data for different climatic zones of the Earth (A), and relative biota productivity (Y/Y_{max}) as a function of dryness index (B): (A) 1, theory; 2, experiment; (B) 1, experiment; 2, averaged curve.

Some examples of thematic maps obtained by means of microwave remote sensing are given in Figures 5.19 to 5.22.

Dryness index

Passive microwaves carried by spacecraft can also provide operational measurements of some generalized parameters of the soil/vegetation system (Shutko and Reutov, 1987). One of these parameters is the 'dryness index', D, which measures the ratio between incident energy and energy used to evaporate moisture contained by the soil. It has been shown theoretically that reductions of T_B and D expansions are more or less similar and are functions of averaged soil temperature and moisture content. Experiments conducted on a world-wide scale from Soviet satellites equipped with microwave radiometers have shown a satisfactory level of agreement between measured and predicted data (Figure 5.23). If one takes into account the fact that D is an indicator of potential plant productivity, and recognizes that operational direct measurements of a 'dryness index' is a complex problem, it becomes evident that microwave radiometry may become an operational instrument in studies of this sort, including energy budget studies.

Acknowledgements

I would like to thank my colleagues, Dr E.A. Reutov, Dr A.A. Chukhlantsev, Dr A.G. Grankov, Dr B.M. Liberman, Eng. Z.G. Jazerian, Eng. A.A. Milshin, Eng. A.A. Haldin and Eng. S.P. Golovachev for their active and fruitful participation in theoretical and experimental investigations in many areas of the world. My sincere thanks are also extended to the organizers of the TERRA-1 conference, especially to Professor Paul M. Mather of the University of Nottingham, for their financial support which gave me the opportunity to participate in the conference and present this paper.

References

Armand, N.A., Reutov, E.A. and Shutko, A.M., 1989, Microwave radiometry of saline soils, in Pampaloni, P. (Ed.), *Microwave Radiometry and Remote Sensing Applications*, Florence, Utrecht: Zeist, VSP, 19-28.

Basharinov, A.E. and Shutko, A.M., 1975, *Simulation studies of the SHF radiation characteristics of soil under moist conditions*, in NASA Technical Translation, F-16, Greenbelt, Maryland, USA: NASA/Goddard Spaceflight Center, 489-510.

Chukhlantsev, A.A., Golovachov, S.P. and Shutko, A.M., 1989, Experimental study of vegetable canopy microwave emission, *Advances in Space Research*, **9**, No. 1, (1)317-(1)321.

Grankov, A.G., Liberman, B.M. and Milshin, A.A., 1989, Experimental studies of radiation characteristics of Earth's covers in microwave, infrared and visible bands, *Investigations of the Earth From Space*, **5**, 98-105 (in Russian).

Kirdiashev, K.P., Chukhlantsev, A.A. and Shutko, A.M., 1979, Land surface microwave radiation under presence of vegetation, *Radioengineering and Electronics*, **24**, 2, 256-61 (in Russian).

Mkrtchjan, F.A., Reutov, E.A. and Shutko, A.M., 1988, Microcomputer-based radiometric data acquisition and processing system for large-area mapping of soil moisture content in the top one metre layer, in *Proceedings of IGARSS'88 Symposium*, Edinburgh, Noordwijk, The Netherlands: ESA Publication Division, 1563-4.

Reutov, E.A. and Shutko, A.M., 1986, Prior knowledge-based soil moisture determination by microwave radiometry, *Soviet Journal of Remote Sensing*, **5**, 100-25.

Reutov, E.A. and Shutko, A.M., 1990, Microwave spectrometry of water content of nonuniformly moistened soil with a surface transition layer, *Soviet Journal of Remote Sensing*, **6**, 72-9.

Shutko, A.M., 1982, Microwave radiometry of lands under natural and artificial moistening, *IEEE Transactions on Geoscience and Remote Sensing*, **GE-20**, 18-26.

Shutko, A.M., 1985, Radiometry for farmers, *Science in the USSR*, **6**, 97-102.

Shutko, A.M., 1986, *Microwave radiometry of water surface and grounds*, Moscow: Nauka/ Science (in Russian).

Shutko, A.M., 1987, Remote sensing of the waters and lands via microwave radiometry (The principles of method, problems, feasible for solving, economic use), in V. Canuto (ed.), *Proceedings of Study Week on Remote Sensing and its Impact on Developing Countries*, Pontifical Academy of Sciences, Vatican City: Vatican City, 413441.

Shutko, A.M., 1990, Institute of Geoinformatics (IGI), Non-Governmental Center for Research, USSR, and Institute of Radioengineering and Electronics (IRE), AS USSR. *Offer on hardware, software and services on survey of soil, vegetation, and water areas from aircraft*, Moscow: IGI, IRE.

Shutko, A.M. and Chukhlantsev, A.A., 1982, Microwave radiation peculiarities of vegetative cover, *IEEE Transactions on Geoscience and Remote Sensing*, **GE-20**, 27-9.

Shutko, A.M. and Reutov, E.A., (1987) On interconnection between radiation field and generalized parameters of natural objects, in Abstracts, International Symposium, Commission F URSI on *Natural Signatures in Remote Sensing*, Göteborg, Sweden: Chalmers University of Technology, 120-1.

Chapter 6
Synergistic use of multispectral satellite data for monitoring arid lands

B.J. CHOUDHURY[1] and
S.E. NICHOLSON[2]

[1]*Hydrological Sciences Branch,*
NASA/Goddard Space Flight Center,
Greenbelt, MD 20771, USA

[2]*Department of Meteorology,*
Florida State University,
Tallahassee, FL 32306, USA

Abstract

Multispectral satellite data, when properly calibrated and standardized, can be used synergistically for a quantitative analysis of land surface change. Relationships among multispectral satellite data (visible reflectance, surface temperature and polarization difference of microwave emission at 37 GHz frequency) have been used to develop hypotheses concerning the relative sensitivity of these data to varied land surface characteristics, which need to be validated by field observations. Radiative transfer models have also been developed to understand these multispectral data. Inter-annual variations of visible reflectance and polarization difference for the period 1982 to 1986 over the Sahel and the Sudan zones of Africa show a lagged response with respect to the rainfall deficit during recovery from drought, which needs to understood in terms of biophysical parameters. Changes in the NOAA satellite orbit (that is, the time of observation) have introduced significant inter-annual variation of infrared surface temperature and have also affected the reflectances to some extent (particularly over arid and semi-arid regions) because of changes in the shadows cast by the vegetation. Some changes in the observations have also occurred because of differences in the response characteristics of the AVHRR sensor on board different NOAA satellites. These sensor/satellite related changes of visible and near-infrared reflectances and surface temperature have to be removed before these data can be used quantitatively to study land surface change.

Introduction

Arid lands are often defined in terms of long-term mean annual rainfall. Evenari *et al.* (1986) suggest that deserts are regions receiving less than 120 mm rainfall, while semi-deserts (steppe) receive up to 400 mm. A definition by rainfall amount is simple to obtain, and for most arid regions rainfall is the only available data. It is, however, difficult to arrive at a single universally accepted definition of deserts because of the existence of a continual change between deserts and steppe and the role of insolation in determining the aridity. The Sahel zone of Africa, defined as the region receiving rainfall between 100 and 400 mm per annum (Nicholson, 1985), has been discussed both within the desert ecosystems (Evenari *et al.*, 1985) and within the savannah ecosystems (Bourliere, 1983). Although stability and resilience (for long-term occupancy and the ability to regenerate on a given site) have been used to describe arid ecosystems void of human intervention, they can become unstable under use (Schlesinger *et al.*, 1990). Anthropogenic pressure is continually increasing over most of the arid lands. Degradation of arid lands over large areas can affect atmospheric conditions in ways which promote further degradation.

Multispectral satellite data, when properly calibrated and standardized, can be used quantitatively to monitor land surface change. Satellite observations in different spectral bands from visible to microwave regions of the electromagnetic spectrum can provide information about the radiative characteristics of the surface, from which biophysical characteristics (such as the fractional ground cover or the vegetation type) can be derived (Choudhury, 1991). Observations of visible and near-infrared reflectances and microwave emission (measured by radiometers) or reflection (measured by radars) can provide information about vegetation characteristics. Different vegetation characteristics (such as chlorophyll or water content of leaves and woody stem area index) exert dominance in different spectral regions. Infrared temperature observations can provide information about surface heat balance. Here, one should recognize that changes of surface characteristics almost invariably affect the heat balance and hence the surface temperature, although the time of observation and atmospheric characteristics also affect the temperature. Microwave observations at longer wavelengths can provide information about surface soil moisture, which is an integrated result of runoff and surface heat balance. Again, changes of surface conditions can affect runoff and consequently the soil moisture. Thus, one can use multispectral observations synergistically to study land surface change.

In this paper we present several consecutive years of satellite and rainfall observations over the sub-Saharan Africa, and discuss these observations in the context of observed rainfall, radiative transfer and heat balance models.

Satellite data

The scanning multichannel microwave radiometer (SMMR) on board the Nimbus-7 satellite was a ten-channel, five-frequency, conically scanning radiometric system which provided global observations at an incidence angle of 50.2°. The equator crossings were at noon and midnight local solar time. The SMMR observations are available from November 1978 to August 1987. Since July 1987 an operational microwave sensor, the special sensor microwave imager (SSM/I), has provided global observations. The SSM/I on board the DMSP-F8 satellite is a seven-channel, four-frequency, radiometric system providing observations at an incidence angle of 53.1°. The equator crossings of the SSM/I are at 0615 and 1815 h local solar time. A second SSM/I was put into orbit at the end of 1990, and observations by both SSM/I are currently available. Both radiometric systems (SMMR and SSM/I) have provided coincident observations for horizontally and vertically polarized brightness temperatures at 37 GHz frequency at a spatial resolution of about 28 km. From these brightness temperatures we have computed the polarization difference, PD (the difference of vertically and horizontally polarized brightness temperatures). Details of the SMMR data processing may be found in Choudhury (1988). The SMMR provided four or five observations at 37 GHz per month at any location from either noon or midnight orbits due to its designated 50 per cent operation time. The SSM/I with full operation time is providing observations at significantly higher temporal frequency (daily or every second day) to allow improved phenological studies.

The daily global area coverage (GAC) observations at a nominal nadir resolution of about 4 km by the advanced very high resolution radiometer (AVHRR) on board the NOAA-7 and -9 satellites have been used to compute visible and near-infrared reflectances and surface temperature for the periods 1982-84 and 1985-87, respectively. The GAC data within the scan angle of 30° from the nadir have been used, and monthly compositing has been done to minimize the effect of cloud and aerosols (Holben, 1986). The calibration constants for converting the digital counts into reflectances take into account the sensor degradation (Holben *et al.*, 1990). The surface temperature has been computed according to the split window technique (Price, 1984; Becker and Li, 1990) taking into account the non-linearity of the sensor response and assuming an emissivity value of 0.96 (no spectral variation). These reflectances and surface temperatures have been registered to the SMMR and SSM/I data.

The spectral reflectances computed after monthly compositing need to be corrected for any cloud-contaminated data remained after compositing, atmospheric aerosols and water vapour. Seasonal and inter-annual variations of aerosols and water vapour are particularly important considerations when evaluating land surface change using the reflectances. The sensor response functions for visible, near-infrared and infrared channels of the AVHRR on board the NOAA-7 and NOAA-9 satellites are somewhat different, leading to small differences in the observations from these satellites. A further complication is the continuously changing time of observations by the AVHRR due to changes in the NOAA satellite orbit. The effect of this changing time of observations is clearly seen in the computed temperatures for surfaces with strong diurnal variations (Table 6.1). Atmospheric water vapour and cloud also affect the microwave data, but generally to a lesser extent. These complicating factors in interpreting the satellite data for land surface change have not yet been resolved satisfactorily.

Modelling: developing a physical basis for synergism

Extensive field observations over agricultural crops and grasses, and radiative transfer simulations for homogeneous leaf canopies, have provided a fairly good understanding of the factors determining visible and near-infrared reflectances. The visible reflectance of green vegetation decreases (due to chlorophyll absorption), while the near-infrared reflectance increases (due to

Table 6.1 Annual average values of visible reflectance (R, in percent) and surface temperature (T, in K) from the AVHRR data and 37 GHz polarization difference (PD, in K) from the SMMR data over rainforest ($0-4°N$, $11-15°E$) and desert ($19-20°N$, $2-12°W$), for different years. The spatial standard deviations are given in parentheses. Any significant inter-annual variation of satellite data over these surfaces may not be attributed to land surface change, however, note the changing surface temperature over the desert.

Year	Rainforest			Desert		
	R	T	PD	R	T	PD
1979	—	—	4.6	—	—	30.0
			(0.3)			(1.6)
1980	—	—	4.4	—	—	30.1
			(0.3)			(1.9)
1981	—	—	4.4	—	—	30.5
			(0.3)			(1.9)
1982	7.6	304.7	4.6	41.8	310.0	30.5
	(0.7)	(0.9)	(0.3)	(1.5)	(1.4)	(1.7)
1983	7.8	304.9	4.4	41.4	309.0	30.7
	(0.8)	(1.1)	(0.3)	(1.3)	(1.7)	(1.8)
1984	7.8	304.1	4.5	41.4	306.9	30.5
	(0.4)	(1.1)	(0.3)	(1.1)	(1.3)	(2.0)
1985	7.2	305.2	4.3	41.5	307.5	29.8
	(0.5)	(1.2)	(0.3)	(1.2)	(1.5)	(1.9)
1986	7.2	304.9	4.5	42.0	309.0	30.3
	(1.0)	(1.2)	(0.3)	(1.2)	(1.4)	(1.7)
1987	6.9	305.4	—	41.5	308.9	—
	(0.5)	(1.1)		(1.3)	(1.8)	

scattering of radiation at the cell wall boundaries within the leaf) as the fractional vegetation cover or the leaf area index increases. These reflectances are also affected by soil reflectance, leaf angle distribution and solar zenith angle. Similar field observations and radiative transfer simulations for microwave emission at 37 GHz frequency show that the PD values for bare soils are affected by soil moisture and surface roughness, and vegetation cover has the effect of decreasing the PD values below those observed over bare soils. The water content of leaves plays a highly significant role in determining polarized microwave emission over vegetated soils (Choudhury, 1990; Choudhury *et al.*, 1990).

Radiative transfer simulations show that an almost linear relation can exist between the visible reflectance and the PD for homogeneous green leaf canopies, varying in the leaf area index (Choudhury, 1990). Such correlations between the visible reflectance and the PD value can exist when photosynthetic integrity (or light absorption characteristics) of chlorophylls is correlated with the turgidity (or the water content) of leaves. However, such correlations may not exist throughout the ontogeny for any crop, for example during certain stages of soil water stress or senescence. Also, changes in the soil reflectance directly affect visible reflectance, but the PD values will be affected when the soil surface roughness or soil surface moisture changes. A direct relation between the PD and the near-infrared reflectance is not expected because the key mechanism responsible for the near-infrared reflectance (namely, scattering of radiation within the leaf at the cell wall boundaries) does not have a counterpart in the microwave emission. Compared to visible and near-infrared reflectances, much less field data is available for 37 GHz emission. Controlled field observations are highly desirable to develop confidence in any model simulation.

The reflectances observed over arid rangelands, shrublands and forests do not always behave as those observed over crops (Smith and Choudhury, 1990a,b; Choudhury, 1991). While the visible reflectance generally decreases with increasing fractional vegetation cover or the leaf area index, the corresponding near-infrared reflectance does not always increase. The factors contributing to these differing responses from crops or grasses include clustering of leaves, high absorptivity of structural (woody) tissues for near-infrared radiation, shadows and the high reflectivity of soils found in arid regions. Radiative transfer models developed for spatially uniform and homogeneous green leaf canopies (the one-dimensional models) need to be extended for understanding the observed reflectances over different plant communities (Myneni *et al.*, 1990).

Simulations using one-dimensional radiative transfer models show that both the visible reflectance and the PD are affected by woody stem area index, but not identically. In most cases the PD decreases more than the visible reflectance as the stem area index increases. Also, the simulations suggest that the effect of spatial heterogeneity of vegetation (clumped foliage canopies as representative of arid regions) should be rather similar for visible reflectance and the PD (Choudhury, 1990). These simulation results need to be further evaluated using three-dimensional models for spatially heterogeneous canopies and validated by field observations.

Surface temperatures derived from infrared observations may be used as an indicator of the surface heat balance (Choudhury, 1991). For non-evaporating surfaces (such as dry bare soils, litter or a stand of senesced vegetation), the surface temperature will generally decrease with increasing albedo (or the average of visible and near-infrared reflectances). The decrease in net radiation caused by the higher albedo is essentially responsible for lower surface temperature, although the actual change in temperature will depend upon soil heat flux and the surface aerodynamic roughness (as determined by surface litter or senesced vegetation). Under similar atmospheric conditions a bare soil will generally be at a higher temperature than that of a stand of senesced vegetation, because a bare soil exchanges heat with the atmosphere less efficiently than a stand of vegetation. The soil heat flux for arid regions could account for as much as 50 per cent of the net radiation for bare soils, but it will be a much lower fraction of the net radiation for vegetated areas. The actual change in surface temperature caused by changes in surface characteristics (for example, land degradation) needs to be understood via heat balance analysis. In contrast to the case for non-evaporating surfaces, the temperature of evaporating surfaces (a stand of green vegetation) decreases as the albedo decreases. In this case, a lower albedo can be an indicator of denser vegetation or wet soil which will be evaporating at a higher rate and therefore will be at a lower temperature than dry soils. These highly contrasting relationships between surface temperature and albedo have been observed in arid regions (Wendler and Eaton, 1983; Menenti *et al.*, 1988). Any change in the fractional vegetation cover (green or senesced) will generally result in a change of surface heat balance, although changes in the atmospheric conditions need to be taken into account for an accurate interpretation of surface temperature in terms of land surface change. Also, different surfaces (sand, soil, senesced and green vegetation) have different infrared emissivities, which have to be considered in interpreting infrared temperature observations. Thus, for example, one needs to recognize that both the surface temperature and the infrared emissivity can change seasonally with vegetation growth and senescence (annual grasslands like the Sahel zone of Africa is an example). Very little information is currently available regarding the spectral variation of infrared emissivity of different surfaces, although such information is needed in order to calculate surface temperature from AVHRR observations (Becker and Li, 1990).

It is clear from the above discussion that one can use multispectral observations synergistically on a rational basis to study land surface change. It is also clear that observations in any spectral band are determined by several land surface characteristics, in addition to the angle of observation and the atmospheric conditions. The relationships between the spectral observations and the surface characteristics are generally non-linear, and it is difficult to

rank the sensitivity of any spectral measurement to different surface characteristics. Relationships between different spectral observations can be used to
develop hypotheses regarding the key surface parameters determining these
spectral data. These would have to be verified by controlled field
observations.

Multispectral observations and relationships

Linear relationships have been observed between visible reflectance and
surface temperature, and between visible reflectance and the PD (Smith and
Choudhury, 1990b; Choudhury, 1990). These spectral data have varied
sensitivity to a number of surface parameters, and, while some of these
parameters are common for these data, it is difficult to determine *a priori* the
rank order of this sensitivity. An hypothesis to explain the observed linear
relations is that the fractional vegetation cover is a primary characteristic
determining these data, and one can postulate the following set of linear
models (Choudhury, 1991):

$$T = fT_v + (1-f)T_s$$
$$R = fR_v(1-f)R_s$$
$$PD = fPD_v + (1-f)PD_s$$

where f is the fractional vegetation cover, T , R , and PD are pixel values of
surface temperature, visible reflectance and 37 GHz polarization difference.
The subscripts v and s refer to the vegetated and bare fractions of a pixel.
These models are approximate, since they do not address the variabilities
within the vegetated and non-vegetated fractions or the observed scatter in
the data values (Kustas *et al.*, 1990). These relations, when validated by field
observations, might be useful for estimating surface characteristics.

Inter-annual variations of R_v and PD over the Sahel and the Sudan zones
of Africa are shown in Figure 6.1, together with the rainfall departure from
the long-term mean rainfall. The rainfall departure has been calculated as the
average value of the departures for rainfall stations within these zones. With
the understanding that an increasing rainfall deficit might indicate progressively less vegetation growth, we see that both R_v and PD increase as the
deficit increases. However, while the rainfall deficit was maximum during
1984, both R_v and PD reached their maximum values during 1985 over the
Sudan zone. A similar pattern for the satellite data is seen for the Sahel zone.
The responses of these satellite data differ with respect to the rainfall deficit,
before and after the severe drought during 1984. These satellite data suggest a
lagged response of surface characteristics with respect to the 1984 drought
(these surface characteristics could be the amount of litter, senesced and/or
green vegetation). Although field observations do show that many biological processes have a lagged response to environmental changes and that these

responses are different for herbaceous (Sahel) and woody (Sudan) vegetation, we need to analyse these data further to understand the factors contributing to this lagged response. One such analysis could assess the response of these satellite data for several years following the 1984 drought. To perform such an analysis, however, we would need to use the SSM/I data because the SMMR time series ended in August 1987.

Much care needs to be exercised in combining observations from different sensors to develop an extended time series for evaluating land surface change. The SMMR and SSM/I observations have been acquired at different times during a day (different equator crossing times), at slightly different spatial resolution and at different incident angle of radiometers. The antenna gain patterns for these two radiometric systems are also not identical. The SSM/I brightness temperatures have been determined with better accuracy than those from SMMR at present, due to experience gained and better technology.

Global observations by both SMMR and SSM/I are available for July and August 1987. A preliminary comparative analysis of the 37 GHz PD values from the SSM/I and the SMMR data is shown in Figure 6.2. Although these PD values are found to be highly correlated, a more thorough comparison of SMMR and SSM/I is being performed to assess any systematic biases

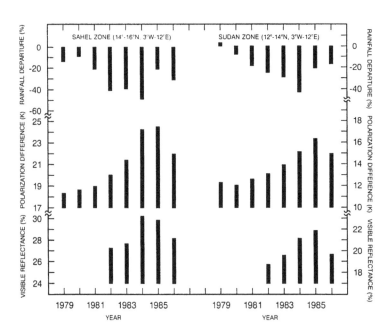

Figure 6.1   Calculated variations of annual rainfall deficit (departure from long-term mean rainfall), visible reflectance (from AVHRR) and 37 GHz polarization difference (from SMMR) over the Sahel and the Sudan zones of Africa. Note the lagged response of both satellite data with respect to maximum rainfall deficit, particularly over the Sudan zone.

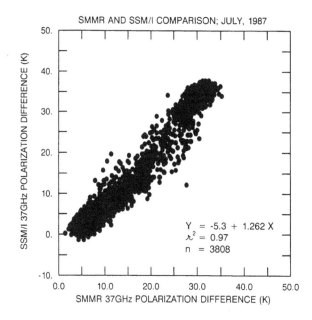

Figure 6.2 Scatterplot of 37 GHz polarization difference for July 1987 calculated from observations by the special sensor microwave imager (SSM/I) on board the DMSP-F8 satellite and scanning multichannel microwave radiometer (SMMR) on board the Nimbus-7 satellite.

associated with any geographic location or vegetation community (or biome). Radiative transfer analysis is also being performed to compare these observations, because of the differences in the data acquisition by the two different sensors. Field and aircraft observations have also been planned to complement modelling and to provide a better understanding of these data.

Summary

More than a decade-long sequence of global observations of spectral reflectances, surface temperature and microwave emission are available for synergistic evaluation of land surface change, particularly over arid and semi-arid regions. Observed inter-annual variations of visible reflectance and polarization difference over the Sahel and Sudan zones have been discussed in relation to rainfall variations. Understanding of these inter-annual variations of the satellite data in terms of biophysical factors needs to be developed through further data analysis together with modelling and field observations.

Acknowledgement

Data processing assistance was provided by Gene Major.

References

Becker, F. and Li, Z.-L., 1990, Towards a local split window method over land surface, *International Journal of Remote Sensing*, **11**, 369–93.

Bourliere, F. (Ed.), 1983, *Tropical Savannas*, New York: Elsevier.

Choudhury, B.J., 1988, Microwave vegetation index: a new long term global data set for biospheric studies, *International Journal of Remote Sensing*, **9**, 185–6.

Choudhury, B.J., 1990, A comparative analysis of satellite-observed visible reflectance and 37 GHz polarization difference to assess land surface change over the Sahel zone, 1982–1986, *Climatic Change*, **17**, 193–208.

Choudhury, B.J., 1991, Multispectral satellite data in the context of land surface heat balance, *Reviews of Geophysics*, **29**, 217–236.

Choudhury, B.J., Wang, J.R., Hsu, A.Y. and Chien, Y.L., 1990, Simulated and observed 37 GHz emission over Africa, *International Journal of Remote Sensing*, **11**, 1837–68.

Evenari, M., Noy-Meir, I. and Goodall, D.W. (Eds), 1985, *Hot deserts and arid shrublands*, New York: Elsevier.

Holben, B.N., 1986, Characteristics of maximum-value composite images from temporal AVHRR Data, *International Journal of Remote Sensing*, **7**, 1417–34.

Holben, B.N., Kaufman, Y. and Kendall, J.D., 1990, NOAA-11 AVHRR visible and near-IR inflight calibration, *International Journal of Remote Sensing*, **11**, 1511–19.

Kustas, W.P., Choudhury, B.J., Inoue, Y., Pinter, P.J., Moran, M.S., Jackson, R.D. and Reginato, R.J., 1990, Ground and aircraft infrared observations over a partially vegetated area, *International Journal of Remote Sensing*, **11**, 409–27.

Menenti, M., Bastiaassen, W., van Eyck, D. and Abd el Karim, M.A., 1988, Linear relationships between surface reflectance and temperature and their application to map actual evaporation of ground water, *Advances in Space Research*, **9**, 165–76.

Myneni, R.B., Asrar, G. and Gerstl, S.A.W., 1990, Radiative transfer in three dimensional leaf canopies, *Transport Theory and Statistical Physics*, **19**, 205–50.

Nicholson, S.E., 1985, Sub-saharan rainfall, 1981–1984. *Journal of Climate Applied Meteorology*, **24**, 1388–91.

Price, J.C., 1984, Land surface temperature measurements from the split window channels of the NOAA-7 Advanced Very High Resolution Radiometer, *Journal of Geophysical Research*, **89**, 7231–7.

Schlesinger, W.H., Reynolds, J.F., Cunningham, G.L., Huenneke, L.F., Jarrell, W.M., Virginia, R.A. and Whitford, W.G., 1990, Biological feedbacks in global desertification, *Science*, **247**, 1043–8.

Smith, R.C.G. and Choudhury, B.J., 1990a, Relationship of multispectral satellite data to land surface evaporation from the Australian Continent, *International Journal of Remote Sensing*, **11**, 2069–88.

Smith, R.C.G. and Choudhury, B.J., 1990b, On the correlation of indices of vegetation and surface temperature over South-Eastern Australia, *International Journal of Remote Sensing*, **11**, 2113–20.

Wendler, G. and Eaton, F., 1983, On the desertification of the Sahel zone, *Climatic Change*, **5**, 365–80.

Chapter 7

Remote sensing inputs to climate models

P.R. ROWNTREE

Hadley Centre,
Meteorological Office,
Bracknell,
Berkshire

Introduction

In this paper the data needed for climate models are indicated, the type of data that remote sensing can be expected to provide is considered, and the extent to which remote sensing has so far provided them is surveyed. In the context of this volume, it is appropriate if the necessity of land surface data is emphasized most.

Climate models require data for several purposes:

(1) data prescribed by the models, for example, for present models, land surface albedo,
(2) to monitor changes in forcing, such as increases in trace gases, or increases in land albedos due to vegetation changes,
(3) for validation of the model simulations, and
(4) for development of parameterization schemes, including validation data and also those inputs to analyses of field data such as the remotely-sensed rainfall data needed to close field experiment water budgets.

Climate models

Before the extent to which remote sensing may provide, and has provided, such data is discussed, it will be useful to explain the structure of the land surface and atmosphere components of a climate model, with particular reference to those parts relevant to land surface processes.

An atmosphere model represents the dynamics and thermodynamics of the atmosphere in terms of the relevant mathematical equations. These are the equations of motion, used to predict changes in the winds on the model's three-dimensional grid, and the equations for prediction of temperature and humidity in three dimensions and the two-dimensional field of total mass or surface pressure.

The grid used to represent these variables typically has a horizontal mesh of about 300 km. However, by the end of the decade, one may anticipate that improvements in computing power will have allowed a reduction of grid mesh to about 100 km. Forecast models, which can be identical in structure, already operate at these and smaller grid meshes. In the vertical dimension the model divides the atmosphere into about 10 to 20 layers. In terms of mass, the layers are thinner near the ground, to allow better representation of the detail of the boundary layer structure. Thus the bottom layer may have a depth of only about 1 per cent of the mass of the atmosphere, or, in height, about 80 m.

The changes in these variables on the gridscale are affected not only by their distribution on that scale, but also by processes on smaller scales. Examples of these are the clouds which modulate the radiative heating, the boundary layer eddies responsible for transferring heat, moisture and momentum from and to the surface, and convective motions which take the heat and moisture from the boundary layer to the 'free' atmosphere above. The effects of these sub-gridscale processes must be represented (or parameterized), in a more or less approximate form, in terms of the larger scale variables.

It is in the representation of these processes that remote sensing can play a role, so it is relevant to note that the magnitudes of the boundary layer and convective energy fluxes are of similar order (about 100 W m^{-2}) to those of the main forcing terms (solar heating of the surface, long-wave cooling of the atmosphere). Even where the typical global figures are smaller, as with the reflection of solar radiation and the sensible heat fluxes, in specific regions—snow and bright sand for the former, dry ground for the latter— these fluxes approach or exceed 100 W m^{-2}.

Data requirements

Table 7.1 summarizes the data needs of the models. Note that these needs exist because we have not included the processes which determine them. Thus, we need to prescribe the atmospheric concentration of ozone because we have not included the necessary chemical model, whereas another gas, water vapour, is modelled and so need not be prescribed. The same applies to

Table 7.1   Equations and the associated data needs.

Equation	Process	Data needs
Momentum	Surface stress	Roughness length (z0)
	Gravity wave drag	Elevation variance
Temperature	Radiation	Albedo, emissivity, atmospheric composition
	Turbulent flux	z0
	Soil heat flux	Soil properties
Moisture	Evaporation	z0, vegetation, soil properties
	Runoff	Vegetation and soil properties
	Percolation	Soil properties

surface quantities; the snow cover is modelled and so need not be prescribed, as it was in some early GCMs, whereas the albedo depends on vegetation and soil which are not modelled, so it must be specified.

Sensitivities of simulations to prescribed data

It is appropriate at this point to consider the sensitivity of the model simulations to these prescribed data. Three examples will be discussed, one each for radiation, hydrology and momentum.

Albedo

Mintz (1984) shows (in his Figure 6.12) the modelled effect of a decrease in albedo of all land from 0.3 to 0.1 in July from a version of the five-layer Meteorological Office model in experiments run by Carson and Sangster. There are increases of rainfall over land, and compensating decreases over the oceans, as the increase in the intensity of the land heat source drives a stronger summer monsoon circulation, weakening the convergence of low-level air and ascent over oceans and strengthening them over land. Changes exceed 5 mm/day, which may be compared with the global mean precipitation of about 3 mm/day.

Hydrology

Mintz (1984) also shows the effect of an extreme change in the land surface hydrology from all wet to all dry (his Figures 6.3 to 6.5, from experiments by Shukla and Mintz). Rainfall is drastically reduced over land, with only parts of the tropical convergence zones continuing to receive over 2 mm/ day, whereas with the wet surface, most of the continental surface had more than this. In this case the reductions were not compensated by changes over the oceans; there were marked decreases there also, except over the equatorial west Pacific, indicating the role of evaporation over land in the hydrological cycle over the oceans. Temperatures were much higher over the dry land, and summer monsoon circulations more intense.

Momentum

Slingo and Pearson (1987) showed the effect on an atmospheric northern hemisphere winter simulation of introducing the frictional effects on the free atmosphere of the absorption of the momentum flux by gravity waves (commonly referred to as gravity wave drag). The inclusion of this parameterization eliminated a serious excessive westerly bias from forecast and climate models. This flux is parameterized at the land surface in terms of the atmospheric flow and the sub-gridscale variance of the surface elevation.

Meeting the prescribed data requirements

These requirements can be met in two ways. For some data, such as the elevation required in the momentum equation, global data sets exist, for example, the US Navy 1/6° orography data set, and these can be used directly. Of course, they may not have adequate accuracy or resolution for the task and may require enhancement or replacement. For example, the 1/6° data are not really sufficiently detailed to calculate the variance of the orography needed to represent the generation of gravity waves. The variance at all scales greater than about 1 km should be represented.

Most of the data required at the land surface depend on the vegetation and soil types. The precise requirements vary from model to model. For example, the Biosphere Atmosphere Transfer Scheme, designed by Dickinson (Dickinson et al., 1986), has about 20 vegetation- and soil-dependent quantities. The Meteorological Office scheme has eight soil-dependent and eight vegetation-dependent parameters, as listed in Table 7.2.

Global data sets inferred from direct observation exist only for a few of the parameters listed in Table 7.2, such as the albedos. For most of the others, global data have not been collected, so they must be constructed on the basis of data sets on vegetation and soil type, and what is known about the dependence of the particular characteristic on the soil and vegetation variables. Vegetation type data sets have been constructed from atlas data by several authors. In the Meteorological Office we use the data set constructed as the result of a collaborative project between the Meteorological Office and the University of Liverpool by Wilson and Henderson-Sellers (1985). In this data set, primary and secondary vegetation types (at least 25 per cent of the land surface) were specified for each 1° grid area. In constructing the model data sets for the grid of 2.5° × 3.75° we currently use, the values for each component 1° gridbox are specified on the basis of the vegetation type and then averaged in a suitable way—generally linearly, except where this is

Table 7.2   Vegetation- and soil-dependent characteristics in the Meteorological Office climate model.

Vegetation	Soil
Vegetated fraction	Heat capacity
Root depth	Heat conductivity
Canopy water capacity	Saturated hydraulic conductivity (K)
Surface resistance to evaporation	Dependence of K on soil water (exponent)
Roughness length	Wilting point
Snow-free albedo	Saturation point
Deep snow albedo	Critical point (below which evaporation is limited)
Vegetation-dependent infiltration enhancement factor	

inappropriate, as for roughness length. Wilson and Henderson-Sellers (1985) also derived data sets of soil coarseness and porosity from the FAO soil data, from which we derive the values for the parameters listed in Table 7.2.

It is clear that better data sets for the quantities in Table 7.2 may be obtained by remote sensing. Again, this may be best done by direct observation of the parameter, as with the snow-free and deep snow albedos, or by deduction of the vegetation type from a remotely-sensed spectral signal, such as the NDVI (normalized difference vegetation index). This latter task was attempted by Henderson-Sellers *et al.* (1986). Comparison with atlas-derived data is encouraging, but shows there is much scope for improvement.

Most of the effort on direct observation has been devoted to the albedo. Widely differing results have been obtained in some areas (e.g. Becker *et al.*, 1988, Figures 2.1 and 2.6). For example, their maps show that over the western Sahara where Dedieu *et al.* (see Rasool, 1987) report values of around 0.4, Preuss and Geleyn (1980) obtain values less than 0.2. Data in Hummel and Reck (1979) give a similar difference from Dedieu *et al.* over the eastern Sahara. Although Dedieu *et al.*'s results look more realistic in many respects, and are closer to the results from atlas vegetation and ground-based albedo data, there are some surprising features, as shown by the transects of the reflectance at longitude 20°E from 18°N to 4°S latitude (Becker *et al.*, 1988, Figure 2.2). These data show values at 5° to 10°N which are 5 per cent or more lower in January than in September. These data are inconsistent with the ground-based data reported by Oguntoyinbo (1970), which show a smaller seasonal variation. One hypothesis is that the effects of tree shadows with a larger zenith angle in January have not been properly accounted for (Franklin, 1988).

Detection of land surface changes

A major objective of climate modelling is the simulation of the effects of changes in the environment. These effects may be realized as increases in greenhouse gas concentrations, or changes in other prescribed parameters, such as the vegetation type or albedo. Remote sensing clearly has an important role here. An example is the deforestation of the Amazon basin in recent decades. Monitoring of the deforestation using satellite images, such as those available from LANDSAT or SPOT, will allow a more realistic specification of the deforestation to be specified in future experiments than the extreme conditions used for sensitivity studies made to date (e.g. Lean and Warrilow, 1989; Shukla *et al.*, 1990). These studies have shown decreases of about 20 per cent in annual mean rainfall within the deforested areas, with some evidence of consistent changes outside. Future experiments will also aim to define the associated changes in vegetation characteristics more realistically, on the basis of both ground-based field studies (e.g. the Anglo-Brazilian ABRACOS study) and remote sensing.

Data for validation of climate models

Climate models need to be validated against real data. Traditionally, this validation has been against climatological data compiled from ground-based observations, and radiosonde observations of the atmosphere. Table 7.3 shows some of the variables relevant to simulations of the land surface for which validation is required and the data sources used.

Ground-based data have been predominantly used for validating simulations of surface data from climate models. The only real exception to this is in respect of snow cover frequency, where visible reflectance observed from satellites has taken over. Recently, microwave data on snow water content has become available and is giving results encouragingly similar to ground-based data. There are also estimates of the surface solar radiation available. Although satellite data are being used to provide valuable evidence on rainfall over the oceans, the quality of rainfall data over land is generally sufficient for climate model validation purposes.

The top of atmosphere (TOA) data on solar albedo and outgoing longwave radiation have proved of considerable value for validation of models. For example, Gates *et al.* (1990) noted differences between various models and these observations, which could be ascribed tentatively to errors in simulations of cloud cover, snow extent and snow albedos.

Table 7.3   Some climate model data requiring validation.

Variable generally used	Validation source
(a) Surface quantities	
Surface temperature	Screen temperature records
Deep soil temperature	Point data
Soil moisture	Point data (Vinnikov and Yeserkepova, 1991)
Evaporation	Local estimates and heat balance atlases
Runoff	Catchment observation
Snow—water content	Point data
Snow—frequency of cover	Remote sensing
Net solar radiation	Point data and heat balance atlases
Net longwave radiation	Point data (sparse) and heat balance atlases
(b) Other quantities	
TOA solar albedo	ERBE satellite data
TOA outgoing longwave	As solar albedo
Precipitation amount	Point data; satellite data

Data needed for development of parameterizations

The development of climate model parameterizations for the land surface is in part advanced through field experiments such as HAPEX-MOBILHY, FIFE, etc. Such experiments require additional data, some of which are provided by remote sensing. As an example, the forthcoming GEWEX Continental-scale International Project (GCIP), designed to study the water balance of the Mississippi basin in the USA, will lean heavily on radar sensing of rainfall to provide the detailed analyses of the atmospheric inputs to the basin water balance. Part of the reason for choosing this location for the experiment is the establishment, now under way, of the NEXRAD radar network.

A major objective of the project is the development of new parameterizations of evaporation and other surface fluxes allowing for the heterogeneity of surface types, precipitation, etc., over the basin. This will require remote sensing to determine the spatial distributions of the surface vegetation and associated surface characteristics. Because of the relatively small scale of the local parameterization development studies and the availability of extensive ground truth data, an experiment of this type can provide many opportunities for the testing and development of other experimental techniques, including aircraft as well as satellite-borne instruments, for remote sensing of variables such as winds, water vapour, aerosol concentrations, soil moisture and surface temperature.

Summary and conclusions

The data requirements for climate modelling can be divided into the prescribed data needed to initialize the models, the data needed for validating the models and so contributing to their development, and data of value in detecting changes in the land surface and thereby useful in specifying the data for experiments to predict the effects of land surface changes. Whilst some of these have mainly to be met from ground-based data, there are many types of data for which the global coverage offered by satellite data is essential. The role of remotely-sensed data is therefore expected to be of increasing importance, though there are major obstacles still to be overcome.

References

Becker, F., Bolle, H.-J. and Rowntree, P.R., 1988, *The International Satellite Land-Surface Climatology Project*. ISLSCP Report No 10, UNEP.
Dickinson, R.E., Henderson-Sellers, A., Kennedy, P.J. and Wilson, M.F., 1986, *Biosphere-Atmosphere Transfer Scheme (BATS) for the NCAR Community Climate Model*, NCAR Technical Note NCAR/TN-275 + STR.

Franklin, J., 1988, Conference presentation at second ISLSCP results meeting, Niamey.

Gates, W.L., Rowntree, P.R. and Zeng, Q.-C., 1990, Validation of climate models, in Houghton, J.T., Jenkins, G.J. and Ephraums, J.J. (Eds), *Climate Change—the IPCC Scientific Assessment*, Cambridge: Cambridge University Press, pp. 93–130.

Henderson-Sellers, A., Wilson, M.F., Thomas, G. and Dickinson, R.E., 1986, *Current global land-surface data sets for use in climate-related studies*, NCAR Technical Note NCAR/TN-272+STR, Boulder, Colorado.

Hummel, J.R. and Reck, R.A., 1979, A global surface albedo model, *Journal of Applied Meteorology*, **18**, 239–53.

Lean, J. and Warrilow, D.A., 1989, Simulation of the regional impact of Amazon deforestation, *Nature*, **342**, 411–13.

Mintz, Y., 1984, The sensitivity of numerically simulated climates to land surface boundary conditions, in Houghton, J.T. (Ed.), *The Global Climate*, Cambridge: Cambridge University Press, 79–105.

Oguntoyinbo, J.S., 1970, Reflection coefficient of natural vegetation, crops and urban surfaces in Nigeria, *Quarterly Journal of the Royal Meteorological Society*, **96**, 430–41.

Preuss, H. and Geleyn, J.F., 1980, Surface albedos derived from satellite data and their impact on forecast models. *Archiv für Meteorologie, Geophysik und Bioklimatologie*, **29**, 345–56.

Rasool, S.I. (Ed.), 1987, Potential of remote sensing for the study of global change, *Advances in Space Research*, **7**, No. 1.

Shukla, J., Nobre, C. and Sellers, P., 1990, Amazon deforestation and climate change, *Science*, **247**, 1322–5.

Slingo, A. and Pearson, D.W., 1987, A comparison of the impact of an envelope orography and of a parameterisation of orographic gravity-wave drag on model simulations, *Quarterly Journal of the Royal Meteorological Society*, **113**, 847–70.

Vinnikov, K.Ya. and Yeserkepova, I.B., 1991, Soil moisture: empirical data and model results, *Journal of Climate*, **4**, 66–79.

Wilson, M.F. and Henderson-Sellers, A., 1985, A global archive of land cover and soils data for use in general circulation climate models, *Journal of Climatology*, **5**, 119–43.

Chapter 8

Atmosphere–biosphere exchange of CO_2

P.G. JARVIS and J.B. MONCRIEFF

Institute of Ecology and Resource Management,
University of Edinburgh

Abstract

The current rate at which the concentration of atmospheric carbon dioxide is increasing is well known and easily measured. What is less clear is the relative role of terrestrial ecosystems and oceans in sequestering carbon. Limits to our understanding of the global carbon balance prevent us from giving unequivocal answers to such seemingly innocuous questions as: how much carbon dioxide is being taken up by the oceans; how much carbon dioxide is being released from soils and vegetation as a result of natural changes and deforestation? Progress in closing the carbon budget must rely on an interdisciplinary approach which relates and integrates observations made at different spatial and temporal scales. When made in combination with detailed biophysical field experiments such observations can reveal the atmospheric and biophysical variables which control carbon dioxide exchange.

Introduction

The steady state that existed 150 years ago between soils, vegetation and the atmosphere has been substantially perturbed by the transfer of carbon dioxide to the atmosphere both from soils and vegetation, as a result of land use changes ('opening up the west') and from the large-scale burning of fossil fuels. This is evident as a rise in the CO_2 concentration of the atmosphere as shown by the record from Mauna Loa in Hawaii extending back just over 30 years (Figure 8.1). The rate of increase in the CO_2 content of the atmosphere is about 3.2 gigatonnes (Gt) of carbon per annum (equivalent to an increase in air concentration of about 1.8 ppmv per annum) and this represents only about half the known current release of CO_2 through the burning of fossil fuels (5.7 Gt of carbon per annum). The fate of the remainder of this CO_2 released, plus any released in land use changes, is poorly known. We can express the global carbon balance as a statement of mass conservation as:

85

$$\frac{\partial C_a}{\partial t} = E + D \tag{1}$$

where $\partial C_a/\partial t$ is the rate of increase in the atmospheric CO_2 content, E is emission and D is deposition. In pre-industrial times, prior to major land use change and the widespread use of fossil fuels, we would expect that the sources and sinks would largely cancel, giving

$$\frac{\partial C_a}{\partial t} = 0. \tag{2}$$

With the onset of industrialization, this steady state has been perturbed and we can write Equation (2) entirely in terms of perturbations (we are not concerned here with the large sources and sinks which were, and probably still are, features of the unperturbed carbon cycle and are likely to be functioning in much the same way as before):

$$\frac{\partial C_a}{\partial t} = \Delta E - \Delta D \tag{3}$$

where ΔE, ΔD represent perturbations to the steady state.

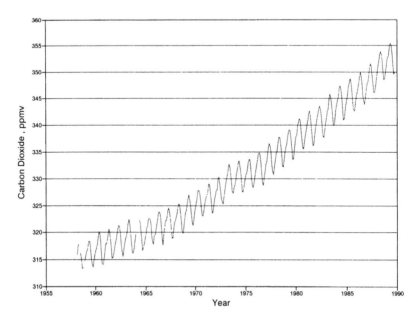

Figure 8.1   Atmospheric CO_2 concentrations measured at Mauna Loa, Hawaii, 1958–89. Plotted from data in Trends '90 (the Carbon Dioxide Information Analysis Center, Oak Ridge, National Laboratory).

Some of the sinks must have taken on an enhanced activity—a CO_2 fertilization effect. The magnitude and spatial distribution of the sinks for anthropogenically-produced CO_2, other than the atmosphere, are uncertain. Data on atmospheric CO_2 are spatially inadequate and the physiological processes governing the responses of plants and soils to the CO_2 fertilization effect is inadequately understood. The major sources arising from land use changes are also poorly known, because of inadequate survey data. Table 8.1 presents a carbon balance sheet for anthropogenically-induced perturbations to the natural carbon cycle. We only know with any degree of certainty the source of CO_2 caused by fossil fuel burning and the size of the atmospheric sink. The other fluxes, marked with a question mark (?) in the table, are highly uncertain, perhaps with errors of the order of ± 100 per cent.

It is also important to state the timescale over which we expect the above equations to be relevant. Over a period of a few months there will be a seasonal influence on carbon assimilation and $\partial C_a/\partial t \neq 0$, even in the absence of major anthropogenic perturbations. On the other hand, in areas of deforestation, the system is likely to be carbon neutral over periods of a few centuries, that is $\partial C_a/\partial t = 0$, because CO_2 released during forest removal will have been taken up again in forest regeneration. An appropriate timescale over which to examine the carbon balance in a world without fossil fuels might be about one to ten years: most soils and vegetation assimilate and respire approximately equal amounts over one year, carbon in grain placed in store will be locked up for perhaps three or four years and in tropical slash and burn agriculture the system will be carbon neutral perhaps over about ten years.

The problem in attempting to close the current carbon balance is then to identify the spatial distribution and magnitude of the anthropogenically-induced sources and sinks. If only half the fossil fuel-derived CO_2 remains in the atmosphere, where are the missing sinks? Once the spatial distribution of sources and sinks has been identified, then direct measurement of net CO_2 exchange and process-based modelling can be used to test specific hypo-

Table 8.1 *The annual global* CO_2 *balance sheet in 1990.*

Sources	Gt/a	Sinks	Gt/a
Fossil fuels[1]	5.7	Atmosphere[4]	3.2
Tropical deforestation[2]	2.1 ?	Oceans[5]	1.0 ?
CO and CH_4 from burning vege-			
tation and soil changes[3]	0.7 ?	Temperate and boreal forests[6]	1.8 ?
		Tropical forests and grasslands[7]	2.5 ?
Total sources	8.5 ?	Total sinks	8.5 ?

Figures are based on: [1,4]Houghton *et al.* (1990); [2]Hammond (1990); Houghton (1991); [3,6,7]Enting and Mansbridge (1991); [5]Tans *et al.* (1990). Entries marked '?' are uncertain.

theses, for instance, relating to the effects of changing climate on sources and sinks, and vice versa. We can, therefore, identify two major areas for discussion: the first approach is based on modelling the observed distribution of CO_2 concentration in the atmosphere (forward and inverse modelling), the second on making direct measurements of CO_2 exchange and coupling these with process-based models (scales of observation). We shall now examine these approaches in turn.

Modelling studies

Inverse modelling

Inverse modelling takes the observed distribution of atmospheric CO_2 concentrations and applies an atmospheric transport model to deduce the space–time dependence of surface sources and sinks of CO_2. The raw data on atmospheric CO_2 concentration can come from one of the three existing networks which regularly sample CO_2 in air using flasks every one to three weeks. Each network has between 20 to 30 sampling stations, some of the stations belonging to more than one network thus enabling inter-comparisons to be made. The peak-to-trough amplitude of the seasonal oscillation varies with latitude and for any station is at a maximum in the late spring/early summer and at a minimum in the autumn. At Point Barrow the peak-to-peak value of the seasonal oscillation is about 18 ppmv, at Mauna Loa (see Figure 8.1) it is about 7 ppmv, and at the South Pole, only about 1 ppmv.

The CO_2 data from each station in the network is analysed for its intra-annual seasonal cycle and its annual trend to establish the temporal variations. The spatial variations of these components and their mean inter-hemispheric gradients are represented by splines (Enting and Mansbridge, 1989, 1991). Inverse modelling tends to amplify any errors in the input data—the observed CO_2 concentration—and, coupled with errors in the transport model, can lead to arbitrarily large errors in the predicted source/sink distributions.

Forward modelling

The forward modelling approach assumes that the geographical distribution of CO_2 sources and sinks is reflected in the spatial and temporal variations of CO_2 concentration in the atmosphere. Forward modelling takes as its starting point some assumed distribution of sources and sinks of CO_2 and attempts to recreate the observed gradients and oscillations (from the flask networks) in atmospheric CO_2 concentration also using a tracer transport model. Fung *et al.* (1983) first proposed using the predicted wind field from a three-dimensional general circulation model (GCM) to advect and convect

CO$_2$ non-interactively, i.e. as a tracer. The GCM had a grid spacing of about 8° latitude and 10° longitude with nine levels in the vertical. The sources and sinks of biosphere CO$_2$ were represented on a 1° × 1° grid spacing. The results from this study were a comparison between the predicted and modelled seasonal CO$_2$ cycle and were reasonably successful. The same tracer model was used later by Tans *et al.* (1990) to simulate the observed atmospheric CO$_2$ concentrations of the NOAA/GMCC flask network. This version of the model also took air–sea transfer of CO$_2$ into account; the partial pressure of CO$_2$ in the surface layers of the ocean, however, are less easily observed than atmospheric CO$_2$ concentrations and this modelling approach will be improved once better observational data is obtained. The tracer model itself has been calibrated using ^{85}Kr and CFC observations, and it was argued that inadequate specification of the biosphere had the largest effect on the precision of the results. The weakness of the forward modelling approach, however, is that any errors in the source distribution of CO$_2$ are suppressed by atmospheric mixing in the model.

Results from the modelling studies

A general conclusion from both forward and inverse modelling concerns the seasonal oscillation in CO$_2$ concentrations which arises because the biosphere is a net sink of atmospheric CO$_2$ during the growing season, as more carbon is absorbed by vegetation than is released by soils, but a net source of atmospheric CO$_2$ at other times of the year when more CO$_2$ is released to the atmosphere than is taken up by plants. The amplitude of this seasonal oscillation is increasing and it is supposed that this may result from either enhanced assimilation of CO$_2$ by vegetation during the growing season or increased decomposition at other times of the year. The decrease in amplitude of the seasonal oscillation from North to South Poles reflects the larger land mass in the northern hemisphere. Perhaps the main conclusion from forward modelling is that in order for the models to recreate the observed north–south atmospheric gradient of CO$_2$, the sinks for CO$_2$ would have to be much larger in the northern hemisphere compared to the southern. As the oceans in both hemispheres have similar partial pressures of CO$_2$ in their surface layers, this implies that the terrestrial biota in the northern hemisphere are more important than previously expected. Inverse modelling work of Enting and Mansbridge (1991) indicates a northern mid-latitude sink with a peak around latitudes 44–53° N (Figure 8.2). Estimates for northern terrestrial biota range from a sink for carbon of between about 2.0 and 3.4 Gt a^{-1}, whereas, following Tans *et al.* (1990), the oceans may sequester less than about 1 Gt a^{-1} carbon. The proposed larger sink in the north is likely to be the result of enhanced plant growth; such enhanced growth may also occur in tropical forests and grasslands and would tend to cancel the carbon release from tropical deforestation. Following Enting and Mansbridge (1991) there is a small net annual source in the tropical equatorial zone between 12°N and 12°S of about 0.6 Gt carbon. This represents the balance between sources, such as the oceans and deforestation, and sinks that are, presumably, the result of the CO_2 fertilization effect. Enting and Mansbridge (1991) argue that the biota in this zone are taking up about 2.5 Gt carbon per annum, as opposed to a deforestation carbon source of about 2.1 Gt a^{-1}. Presently there are few measurements to confirm this major equatorial sink but the work of Fan et al. (1990) indicates that such a sink is possible, based on a very limited data set. Measurements are being made to test these hypotheses and we turn now to discussion of the types and scales of measurements which are being employed.

Scales of observation

Direct measurement of CO_2 fluxes has a longer history than modelling and occurs on many different spatial and temporal scales. If a goal of measurements is to validate predictions from modelling exercises, then it is necessary to resolve and integrate these varied scales of observation (Figure 8.3). Global carbon modelling at present scales of resolution only makes use of a minimal amount of process-based information on the biology of the sources and sinks. It is therefore vital that, in scaling up measurements from one scale to another, only the most essential information on controls are identified and modelled. That is the great challenge of integrating observations over a hierarchy of different temporal and spatial scales. The number of scales also

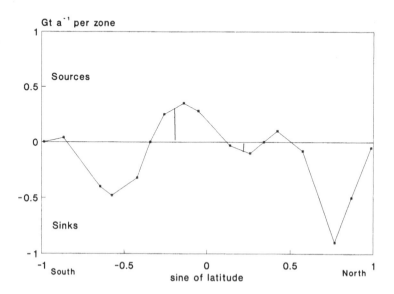

Figure 8.2   Sources and sinks of carbon calculated by the inverse modelling study of Enting and Mansbridge (1991). The short vertical lines are ±12° of the equator.

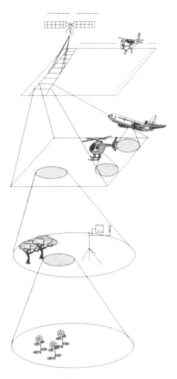

Satellite Radiometry
10 m - 100 km

Regional Flux
1 km - 100 km

Airborne Radiometry
10 m - 15 km

Land Surface Flux
10 m - 1 km

Leaf Physiology
1 cm - 10 m

Figure 8.3 Scales of observation under study in a typical land–atmosphere observational experiment (redrawn from Sellers et al., 1988).

indicates the interdisciplinary nature of this type of research with biologists operating at the smaller scales and global climate modellers at the larger scales.

Leaf scale

The CO_2 and water vapour exchanges of isolated leaves enclosed in small chambers, such as cuvettes and porometers, have been intensively studied in both controlled and natural environmental conditions. The response of individual leaves to imposed conditions of ambient temperature, humidity and CO_2 have resulted in well defined response functions at the leaf scale of a few centimetres and usually a few tens of seconds (e.g. Jarvis and Sandford, 1986). Thus the basic properties of leaves with respect to CO_2 and water vapour fluxes have been well defined and the controls on the fluxes, such as stomata, are adequately understood and can be used in system modelling.

Plot scale

Plants can be grown in plots either to study their response to some short term stimulus or for long term acclimation studies such as their response to high

CO_2 concentrations. Leaves on plants constituting areas of vegetation do not exist in isolation; they respond to atmospheric conditions which are to some extent dependent on the response of all the other leaves in the vicinity (Jarvis and McNaughton, 1986). Individual plants or small areas of vegetation can be isolated in large chambers (a few cubic metres) both for measurement and for conditioning. Individual branches and even mature trees can be fumigated with pollutants and enriched with CO_2 inside such chambers. The influx of CO_2 to isolated parts of plants, entire plants and small areas of vegetation has been widely investigated using such chambers, enabling an integration to be made directly over a large number of leaves. Unfortunately, however, such chambers decouple the leaves from the atmosphere and prevent or distort the feedbacks between the vegetation and the atmosphere that would occur naturally in the absence of the chamber, as well as substantially changing the radiation environment (Leuning and Foster, 1990).

By growing plants in conditions different from the present ones, for periods of a few months or seasons, it is possible to observe the responses of growth and phenology and the resultant acclimation of physiological processes to a changed environment. This approach has been used to show that the impact of aerial fertilization of plants by CO_2 can be substantial. For example, Idso *et al.* (1991) report an experiment in which young orange trees were transplanted to open-top chambers, one group receiving present air whilst the other was fumigated with air which had been CO_2-enriched by about 300 ppmv above the present concentration. After $2\frac{1}{2}$ years the trees in the elevated CO_2 chambers had twice as much leaf and fine root biomass as the trees grown in air of the present CO_2 concentration; the tree trunks and branch volumes were about 2.8 times those of the control trees. The modelling studies discussed earlier suggest a strong CO_2 fertilization effect and this is consistent with such experimental results.

Stand scale

Measurements of carbon fluxes from whole fields of crops or stands of trees can be made by micro-meteorological methods. Such methods fall into two main groups, one based on measuring gradients above the stand of atmospheric properties such as temperature, humidity, CO_2 or wind speed (aerodynamic or Bowen ratio methods) and the other based on directly sampling the air for its concentration of CO_2 and also the magnitude and direction of the vertical component of wind speed (the eddy covariance method). Because of the emerging dominance of the eddy covariance technique, we explain briefly some of the basic principles here and give some indication of current developments.

The vertical flux of carbon dioxide (F_c) can be written simply as

$$F_c = \overline{w'c'} \qquad\qquad (4)$$

where w' is the instantaneous fluctuation in vertical velocity about the mean and c' is the instantaneous fluctuation of CO_2 from the mean (Figure 8.4). The overbar shows that the flux is simply the product of these two fluctuations averaged over a suitable interval, usually of the order of 10 to 30 minutes. Fluctuations are measured at typical rates of 10 to 20 Hz (depending on aerodynamic roughness of the surface and thermal stability). Figure 8.4 shows that during the day, and over this actively growing spruce forest, air which is moving down towards the canopy is relatively rich in CO_2, whereas air which is moving up from the canopy is relatively depleted of CO_2, the CO_2 having been used up in the process of photosynthesis. The net amount of CO_2 passing the sampling point is given by Equation (4).

Infrared gas analysers used in eddy covariance exist in two different configurations. 'Open-path' analysers typically have the source and detector of infrared radiation separated by an open path of 10 to 20 cm through which air is allowed to pass freely; they have response times of the order of 0.1 s . In using 'closed-path' analysers, air is ducted down a narrow inlet tube from the sampling point to the optical bench of the analyser; response times are of the order of a second or more, although in some instruments small optical benches reduce the response times to close to those of open-path systems (Figure 8.5). Open-path analysers are still research tools and there are, at present only two manufacturers world-wide producing them. Closed-path analysers are to some extent unmodified laboratory analysers and consequently have a long history of development and use. So long as they are operated above tall, rough vegetation (or at some greater height above

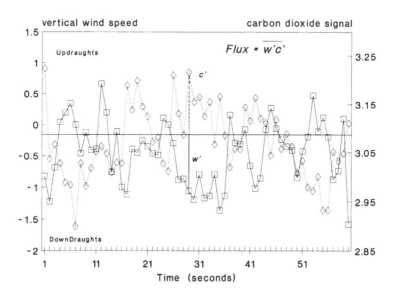

Figure 8.4 Rapid response of CO_2 and vertical wind speed signals measured as part of an eddy covariance experiment. (*Values on the vertical axes are arbitrary.*)

Figure 8.5   Eddy covariance instrumentation exposed above a spruce canopy; a one-component sonic anemometer lies between the inlet tubes for the closed-path analysers and the open-path analyser and a three-component sonic anemometer is in the background.

shorter vegetation) they have the potential to be used in long-term, routine flux monitoring studies.

In use, existing open-path analysers cannot be left unattended for long periods as they are not generally weatherproof, some of the optical lenses used being made of CaF_2 which is damaged by exposure to rain. In addition, one of the commercial analysers available needs to be calibrated regularly over the course of a day as its gain is dependent on instrument temperature. As with all non-dispersive infrared gas analysers, it is the density of CO_2 in air which is measured rather than the mixing ratio or mole fraction, etc., and this is dependent on the density of air. Simultaneous fluxes of heat and water vapour cause the air to expand and hence affect the density of CO_2. Corrections can be made to the raw flux of CO_2 as measured by open- and closed-path analysers (Webb *et al.*, 1980). The correction to the CO_2 flux caused by sensible heat can be as large as 50 per cent of the 'raw' CO_2 flux as measured and of the opposite sign. Corrections for latent heat flux are smaller, usually less than about 10 per cent of the raw CO_2 flux. Open-path analysers require both corrections to be applied and, although the corrections have been well validated by field experiments (for example, Leuning *et al.*, 1982), users tend to be unhappy about arriving at the 'true' flux *via* such large corrections. With closed-path analysers, by contrast, it is not necessary to apply the correction for sensible heat flux because temperature fluctuations are effectively smoothed out in passage down the inlet tube to the optical bench and the air is heated to constant temperature in passing through the analyser (although the smaller, latent heat flux correction remains). The tubing also serves to dampen high frequency fluctuations in the sample airstream, some of which may be responsible for carrying some of the CO_2 flux; corrections can be calculated for this effect (Leuning and Moncrieff, 1990). The contribution to the flux from small, high frequency eddies is very small anyway over rough vegetation and their relative importance diminishes with height above the surface. Figure 8.6 shows results from an intercomparison of three instruments—a commercial open-path analyser and two closed-path analysers. The size of all the corrections applied to either of the closed-path analysers amounted to between 10 and 20 per cent of the 'raw' flux, whereas for the open-path analyser the corrections required were about 50 per cent of the 'raw' flux. The good agreement shown here between the two different types of CO_2 analyser indicates the potential use of the closed-path system for routine and long-term flux monitoring studies.

Regional scale

Eddy covariance measurements are usually made from towers with the sensors located a few metres or so above the vegetation. At that height they

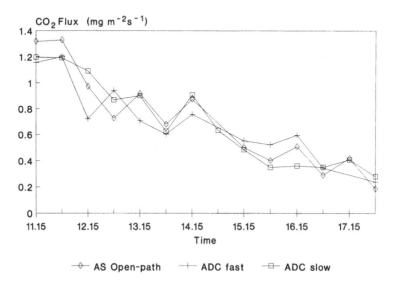

Figure 8.6   An intercomparison of open- and closed-path CO_2 analysers over spruce forest, 25 July 1990, Rivox, SW Scotland. The open-path analyser is marked AS in the figure, the two closed-path analysers are marked 'ADC$_{fast}$' (response time 0.1 s) and 'ADC$_{slow}$' (response time 1 s).

sample air which is representative of perhaps a few hundred metres upwind (depending on atmospheric stability and aerodynamic surface roughness). Eddy covariance instruments can also be mounted on aircraft and flown within the planetary boundary layer, e.g. at altitudes typically up to 1000 m during the day. At these heights the flux is representative of a much larger area on the ground, up to 10×10 km. One problem which continues to trouble experimenters using this technique, however, is that aircraft-derived fluxes consistently underestimate the flux as seen by surface flux stations—an apparent loss of flux, usually in the range 0 to 40 per cent, has been demonstrated in a number of recent land-surface observational studies (Andre et al., 1988; Sellars et al., 1988). The current explanation is that some component of the eddy flux may be at scales which are much longer than the scales that are normally associated with the planetary boundary layer (Shuttleworth, 1991). Until this degree of uncertainty is reduced, airborne flux measurements are likely to be used only to parameterize and validate the predictions from boundary layer modelling studies rather than being relied upon to produce area-averaged fluxes.

A promising technique for studying atmosphere–biosphere exchange of CO_2 involves remote sensing. Leaves absorb much of the visible radiation incident upon them but reflect strongly in the near-infrared. Thus the ratio of red:far-red reflectance reveals much about the photosynthetic ability of the vegetation and the ratio will change with time as the chlorophyll content of the leaves and stand changes. One common index is the 'normalized difference vegetation index' (NDVI) which is a combination of reflectances in the visible C_{VIS} (0.4–0.7 μm) and near-infrared C_{NIR} (0.7–1.1 μm). It is defined as

$$NDVI = (C_{NIR} - C_{VIS})/(C_{NIR} + C_{VIS}).$$

The NDVI for a desert where there are few leaves would typically be $\leqslant 0.02$ whereas for a healthy green crop covering the ground it would be $\geqslant 0.5$. The rate of influx of CO_2 to vegetation from the atmosphere is closely correlated with the rate of increase of dry matter. Thus an index which can sense changes in dry matter content in an area also indirectly has the capacity to reveal rates of CO_2 influx to the surface. As Choudhury and Nicholson (this volume, Chapter 6) point out, however, the NDVI of semi-arid crops is complicated by shadowing and clustering effects which may make this approach less useful in such areas. One interesting application of this technique was demonstrated by Fung et al. (1987) when they used NDVI estimates obtained from the NOAA-7 polar orbiting satellite, in conjunction with field data on soil respiration and certain assumptions about the geographical distribution of major vegetation types, and obtained monthly estimates of the net CO_2 flux to the land surface. These monthly fluxes were then used in a forward modelling study to simulate the seasonal oscillations in atmospheric CO_2 concentrations, as referred to earlier.

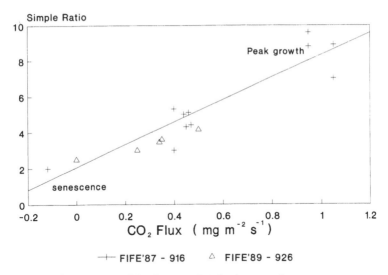

Figure 8.7   Simple ratio versus CO_2 flux, FIFE. The data come from two separate sites in two different years. The 1987 data comprises four experimental periods when the vegetation (prairie grass) was in different stages of growth and was obtained by Professor S.B. Verma of the University of Nebraska; the 1989 data are for a site located about 5 km south of the site used in 1987. The CO_2 fluxes are averages over the period 1100-1400 Local Time and the simple ratio was obtained from a Barnes MMR (Molecular Multispectral Radiometer) mounted on a helicopter.

An alternative reflectance ratio known as the Simple Ratio, is the ratio of infrared (0.75–0.88 μm) to red (0.63–0.68 μm) wavelengths and this ratio has also been found to be closely correlated with the chlorophyll content of vegetation (Sellers, 1985). During FIFE (First ISLSCP Field Experiment— ISLSCP is the International Land Surface Climatology Project—see Shuttleworth, this volume, Chapter 9), the Simple Ratio was obtained from aircraft-borne radiometers and Figure 8.7 shows the relationship obtained with ground-based eddy covariance measurements of CO_2 flux. The potential of this technique to obtain CO_2 fluxes by satellite remote sensing will be tested in other land–atmosphere experiments which are planned. Such ratios may also yield information on fluxes of water vapour from the land surface because of the relationship between the gain of CO_2 and loss of H_2O in photosynthesis.

Conclusions

(1) The proposed CO_2 fertilization effect on terrestrial biota and the role of the oceans are large uncertainties in attempting to close the global carbon budget.
(2) Forward and inverse modelling are essential methods to generate hypotheses concerning sources and sinks of carbon.

(3)   The current flask networks are biased towards oceanic areas and there is a need for additional stations in the north temperate zone, where potentially much of the 'missing sink' might be found.

(4)   The hypotheses generated by modelling should be tested by measurements over the major biomes of the earth, including oceans.

(5)   Eddy covariance is the most promising micrometeorological technique for measuring CO_2 fluxes and developments in closed-path analysers should enable routine and long-term flux measurements to be made.

(6)   Further development and testing of reflectance ratios promise to enable CO_2 flux estimates to be made from satellites as part of their routine observational strategies.

There is some urgency to achieve the above aims as appropriate management strategies are required now to combat global change. The international cooperation of scientists from different disciplines on land surface observational experiments will encourage significant new progress in these areas.

References

Andre, J.-C., Goutorbe, J.-P., Perrier, A., Becker, F., Bessemoulin, P., Bougeault, P., Brunet, Y., Brutsaert, W., Carlson, T., Cuenca, R., Gash, J., Gelpe, J., Hildebrand, P., Lagouard, P., Lloyd, C., Mahrt, L., Mascart, P., Mazaudier, C., Noilhan, J., Ottle, C., Payen, T., Phulpin, T., Stull, R., Shuttleworth, J., Schmugge, T., Taconet, O., Tarrieu, C., Thepenier, R.M., Valencogne, C., Vidal-Madjar, D. and Weill, A., 1988, HAPEX-MOBILHY: First results from the special observing period, Annals of Geophysics, **6**, 477–92.

Enting, I.G. and Mansbridge, J.V., 1989, Seasonal sources and sinks of atmospheric CO_2: direct inversion of filtered data, Tellus, **41B**, 111–26.

Enting, I.G. and Mansbridge, J.V., 1991, Latitudinal distribution of sources and sinks of CO_2: results of an inversion study, Tellus, **43B**, 156–70.

Fan, S.-M., Wofsy, S.C., Bakwin, P.S. and Jacob, D.J., 1990, Atmosphere–biosphere exchange of CO_2 and O_3 in the Central Amazon Forest, Journal of Geophysical Research, **95**, 16851–64.

Fung, I.Y., Prentice, K., Matthews, E., Lerner, J. and Russell, G., 1983, Three-dimensional tracer model study of atmospheric CO_2: response to seasonal exchanges with the terrestrial biosphere, Journal of Geophysical Research, **88C**, 1281–94.

Fung, I.Y., Tucker, C.J. and Prentice, K.C., 1987, Application of very high resolution radiometer vegetation index to study atmosphere-biosphere exchange of CO_2, Journal of Geophysical Research, **92**, 2999–3015.

Hammond, A.L. (Ed.), 1990, World Resources 1990-91, New York: Oxford University Press.

Houghton, J.T., Jenkins, G.J. and Ephraums, J.J. (Eds.), 1990, Climate Change: The IPCC Scientific Assessment, Cambridge University Press.

Houghton, R.A., 1991, Tropical deforestation and atmospheric carbon dioxide, Climate Change, **19,** 99–118.

Idso, S.B., Kimball, B.A. and Allen, S.G., 1991, CO_2 enrichment of sour orange trees: $2\frac{1}{2}$ years into a long term experiment, Plant, Cell and Environment, **14**, 351–3.

Jarvis P.G. and McNaughton, K.G., 1986, Stomatal control of transpiration: scaling up from leaf to region, Advances in Ecological Research, **15**, 1–49.

Jarvis, P.G. and Sandford, A.P., 1986, Temperate forests, in Baker, N.R. and Long, S.P. (Eds), Photosynthesis in Contrasting Environments, Amsterdam: Elsevier, 200–36.

Leuning, R. and Foster, I.J., 1990, Estimation of transpiration by single trees: comparison of a ventilated chamber, leaf energy budgets and a combination equation, *Agricultural and Forest Meteorology*, **51**, 63–86.

Leuning, R. and Moncrieff, J.B., 1990, Eddy covariance CO_2 flux measurements using open- and closed-path CO_2 analysers: corrections for analyser water vapour sensitivity and damping of fluctuations in air sampling tubes, *Boundary Layer Meteorology*, **53**, 63–76

Leuning, R., Denmead, O.T., Lang, A.R.G. and Ohtaki, E., 1982, Effects of heat and water vapour transport on eddy covariance measurement of CO_2 fluxes, *Boundary-Layer Meteorology*, **23**, 209–22.

Sellers, P.J., 1985, Canopy reflectance, photosynthesis and transpiration, *International Journal of Remote Sensing*, **6**, 1335–72

Sellers, P.J., Hall, F.G., Strebel, D.E., Kelly, R.D., Verma, S.B., Markham, B.L., Blad, B.L., Schimmel, D.S., Wang, J.R. and Kanemasu, E., 1988, *First ISLSCP Field Experiment, April 1988 Workshop Report*, Code 624, NASA/Goddard Spaceflight Center, Maryland, USA.

Shuttleworth, J., 1991, Insight from large scale observational studies of land/atmosphere interactions, *Surveys in Geophysics*, **12**, 3–30.

Tans, P.P., Fung, I.Y. and Takahashi, T., 1990, Observational constraints on the global atmospheric CO_2 budget, *Science*, **247**, 1431–8.

Webb, E.K., Pearman, G.I. and Leuning, R., 1980, Correction of flux measurements for density effects due to heat and water vapour transfer, *Quarterly Journal of the Royal Meteorological Society*, **106**, 85–100.

Chapter 9
Observational studies of the land/atmosphere interaction

W.J. SHUTTLEWORTH

Institute of Hydrology,
Wallingford,
Oxfordshire

Abstract

This paper provides an overview of major international studies of the land/atmosphere interaction carried out during the 1980s, with particular attention to ARME, HAPEX-MOBILHY and FIFE, and previews those proposed for the 1990s. Selected results, either particular to individual studies or common to several, are interpreted to provide insight on the value and reliability of experimental data and field systems and to provide guidance for future experiments. Progress in addressing the need to provide area-average aggregate values for surface energy fluxes is assessed, and a recommendation made for enhanced attention to the use of planetary boundary layer development as an indirect measure of these. Proposed observational studies on land/atmosphere interactions under the World Climate Research Programme and the International Geosphere/Biosphere Programme are described.

Introduction

The simple observation that climates over continents differ profoundly from those over oceans at approximately the same latitude demonstrates that the climate responds to the presence of a land surface and the vegetation growing there. This fact, coupled with simple sensitivity studies made with General Circulation Models (GCMs) in which the description of the land/atmosphere interaction is altered (Charney *et al.*, 1977; Shukla and Mintz, 1982; Sud et al., 1985), demonstrates that land/atmosphere interactions affect climate. We can now be unequivocal about this.

This is an extremely important fact. Most of the attention on predicted change consequent on enhanced 'greenhouse warming' has been focused on increases in air temperature, but this perhaps reflects limitations in the models, rather than need for policy formulation. Since man is a land-based species, the primary requirement for realistic response strategies is for prediction of near-surface climate over land surfaces at regional scale, and particularly for prediction of precipitation fields. There can be no real hope of accuracy or credibility in this unless GCMs include adequate descriptions of the land/atmosphere interaction. Perhaps even more obvious than this, climate change in response to major modifications in land use, which now seem inevitable in consequence of projected growth in global population, will necessarily require greatly improved definition of land/atmosphere interactions since this is the very essence of the change involved.

Enlightened portions of the community have foreseen the needs described above and for a decade there have been observational initiatives which sought to refine the capabilities of models to represent terrestrial surfaces. Such studies have arisen in three main ways. Some have been independent, perhaps bilateral, initiatives—and, fortunately, this still occurs to some extent—but most have occurred under the umbrella of international prgrammes. The World Climate Research Programme (WCRP) has an important and still ongoing programme of experiments called the Hydrologic Atmospheric Pilot EXperiments (HAPEX), whose primary objective is to improve the representation of 'hydrology' in General Circulation Models and, in particular, its representation at the area-average scales used in models of the global system. A second series of experiments has proceeded under the International Satellite Land Surface Climatology Project (ISLSCP), which has been primarily fostered through the international space agencies, particularly NASA, and seeks to provide parameterization of global models indirectly from satellite data after validation by 'ground truth' observational studies. In the next section we select the best examples of experiments which have arisen in these three ways, in order to illustrate the differing approaches and the level of achievement.

This paper draws substantially, though selectively, on previously published material (Shuttleworth, 1991a,b,c). The interested reader is referred to these papers for more comprehensive information.

Past experiments

Table 9.1 and Figure 9.1 together show the location of experiments carried out in the 1980s and their relation to international programmes where appropriate. Two major studies of the HAPEX type have taken place, one in South-West France, HAPEX-MOBILHY, and one in China, the HEIFE experiment. There have been several small-scale experiments carried out under the ISLSCP banner, but two major experiments, the First ISLSCP Field Experiment (FIFE) in Kansas, and the LOngitudinal land-surface TRansverse EXperiment (LOTREX) in Germany. The independent initia-

Table 9.1 Summary of past climate prediction-related observational studies.

Location	Experiment	Affiliation	Timing of activity	Scale	Duration
1	Amazon Region Micrometeorology Experiment(ARME)	Independent	1983–1985	Single site	2 years, 4 long IFCs
2	Hydrological Atmospheric Pilot Experiment (HAPEX-MOBILHY)	HAPEX	1986	100×100 km	1 year, 1 long IFC
3	First ISLSCP Field Experiment (FIFE)	ISLSCP	1987, 1989	15×15 km	2 seasons, 5 short IFCs
4	'LaCrau' Experiment	ISLSCP	1987	Multisite	1 season, 1 short IFC
5	Longitudinal land-surface Transverse Experiment (LOTREX-HIBE88)	ISLSCP	1988	10×10 km	1 season, 1 long IFC
6	Joint Greenland Field Experiment	ISLSCP	1988	Multisite	1 season, 1 long IFC
7	'Niger' Experiment	ISLSCP	1988	Multisite	1 season, 1 short IFC
8	Sahelian Energy Balance Experiment (SEBEX)	Independent	1988–1990	Multisite	3 seasons, 3 long IFCs
9	'Botswana' Experiment	ISLSCP	1989	Multisite	1 season, 1 long IFC
10	Heihe River Experiment	HAPEX	1989	70×90 km	1 year, 4 short IFCs
11	'Sudan' Experiment	ISLSCP	1989	Multisite	1 season, 1 short IFC

Notes: 1. The locations are indexed to Figure 9.1;
2. 'Season' refers to an experiment with activity restricted to summer months;
3. An Intensive Field Campaign (IFC) is defined as short if of less than 3 weeks duration.

tive which has gained most attention, meanwhile, was a bilateral Anglo-Brazilian study of the atmospheric interaction of tropical rainforest carried out in Amazonia and called the Amazon Region Micrometeorological Experiment (ARME).

Amazon Region Micrometeorological Experiment (ARME)

The Amazon rainforest is the least disturbed portion of the tropical forest biome and, as such, provides an opportunity to provide at least a first order description of the interaction of this important land cover type through a single site experiment. ARME (Shuttleworth *et al.*, 1984; Shuttleworth, 1988a) was just such a single point study and was mounted in central Amazonia over an area of undisturbed tropical rainforest in a forest reserve some 25 km from the city of Manaus. It comprised extensive micrometeorological measurements carried out on a 45 m high scaffolding tower above 35 m high forest canopy. Hourly measurements of near-surface meteorological variables were maintained on this tower for a period of 25 months, along with measurements of the energy, water vapour and momentum transfer between the forest and the atmosphere during three intensive study periods lasting typically 10 weeks at selected periods of the year. In addition, detailed measurements were made of the processes involved in the forest atmosphere interaction, in particular of the water lost through re-evaporation of the precipitation intercepted on the forest canopy, of the movement and storage of moisture and energy in the soil, and measurements of the differing interactions of leaves at various heights through the canopy.

The results of this study are too numerous for inclusion in a review of this sort, but three important results are selected as of global relevance. Firstly,

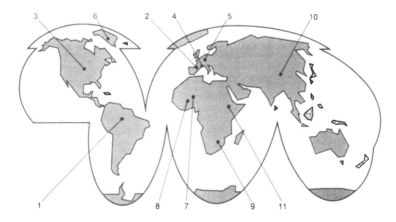

Figure 9.1   The location of the past climate prediction-related observational studies. The numbers here are those of the experiments listed in Table 9.1 (from Shuttleworth, 1991).

the study demonstrated (Figure 9.2) that half of the (very seasonal) precipitation input to the forest at this location was re-evaporated at a more uniform rate through the 25-month study period (Shuttleworth, 1988a). This confirmed the previously hypothetical suggestion that the Amazon forest is to this extent responsible for maintaining its own climate. Water evaporated from areas near the mouth of the river is precipitated further upwind in the basin, and this cycle is repeated until the water reaches the Andes in the east.

A second major result was to demonstrate that some of the models previously used to make predictions of the climatic consequences of Amazonian deforestation had been in significant error. It was easily shown that very simple representations involving no attempt to describe the effect of the vegetation canopy, the so-called Bucket models, gave poor simulation of dry-canopy evaporation rates (Sato *et al.*, 1989). Even sophisticated descriptions of the forest canopy, when applied in GCMs with weaknesses elsewhere, were shown to have provided poor descriptions of the existing tropical forest cover (Figure 9.3) (Dickinson, 1989a,b; Shuttleworth and Dickinson, 1989). However, once the ARME data became available, it was possible to use them to calibrate or validate GCM descriptions (for example, Sellers *et al.*, 1989), and in this way provide successive improvement in the predictive accuracy of such GCMs. Current predictions (e.g. Shukla *et al.*, 1990) are therefore now more credible and predict a 2° rise in temperature, a 20 per cent reduction in evaporation, and an ensuing 30 per cent reduction in precipitation for the Amazonian region (Figure 9.4). It should be emphasized, however, that models still possess no calibration of post-deforestation surfaces, have untested descriptions of cloud cover response to changing

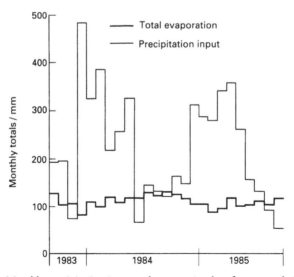

Figure 9.2   Monthly precipitation input and evaporation loss for a central Amazonian site, as measured in ARME (from Shuttleworth, 1988a).

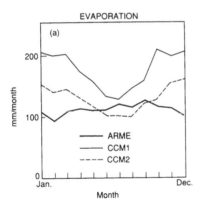
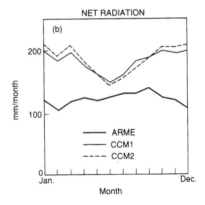

Figure 9.3 Comparison between monthly average (a) evaporation and (b) net radiation as measured at a site near Manaus in central Amazonia, and that modelled in two simulation experiments using the NCAR Community Climate Model (CCM). Differences are mainly due to weakness in the model's ability to simulate convective cloud cover at the time of the simulation experiments (from Dickinson, 1989).

surface conditions, and can make no representation of mixed change in forest cover.

HAPEX-MOBILHY

The HAPEX-MOBILHY experiment (Andre *et al.*, 1988) took place in South-West France in 1986. The primary objective was to investigate the provision of large-scale area-average description of mixed temperate vegetation. The study area comprised a 100×100 km square, including agricultural crops (vineyards, sunflowers, etc.), and a significant area (approximately 40 per cent) of reasonably uniform maritime pine forest. The experiment involved separate measurement and modelling of the interaction of the component vegetation present, and their synthesis through mesoscale meteorological models, to provide the required area–average description. Micrometeorological measurements were therefore made over the forested area and the primary agricultural cover types. In addition there was boundary layer sounding with aircraft and radiosondes, and remote sensing with airborne and space-borne systems; these measurements were taken in order to check the adequacy of the representation of the mesoscale model synthesis.

The experimental data clearly demonstrated the different interaction of the vegetation covers present at the daily timescale (Figure 9.5a). Since the forest and agricultural cover were spatially distributed in a systematic way, this difference in interaction was visible in the airborne measurements

(Figure 9.5b). The difference was present throughout the experiment, but on particular occasions it became visible from space in the form of mesoscale convective clouds due to enhanced convective activity over the forested area (Shuttleworth, 1988b). A very positive feature of the subsequent analysis was the fact that it proved possible to represent these differing interactions in mesoscale meteorological models (Pinty *et al.*, 1989; Mascart *et al.*, 1990). It is significant that such models implicitly provide area-average integration of the surface fluxes and near-surface meteorological variables, thereby potentially providing synthetic data which could be used to optimize surface representations at a (larger) area scale approaching that used in GCMs.

A further suggestion based on the HAPEX-MOBILHY data (Shuttleworth, 1988b) is that the average representation of terrestrial surfaces may respond to the organization present in vegetation cover. The disorganized agricultural crop cover pervading most of the study area seemed adequately

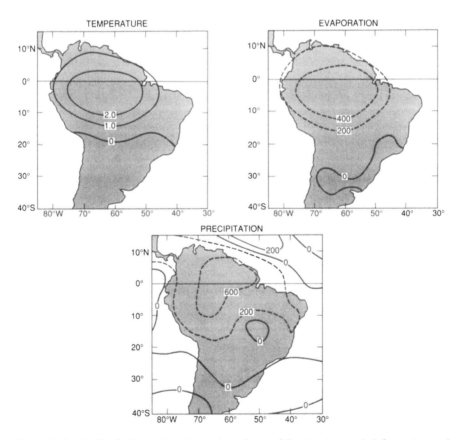

Figure 9.4 Predicted change in Amazonian climate following its total deforestation and replacement with pastureland (from Shukla et al., 1990).

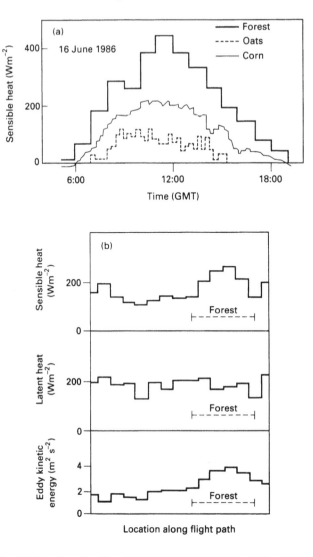

Figure 9.5 Sensible heat flux data from HAPEX-MOBILHY for 16 June 1986, illustrating: (a) differences for selected vegetation covers, from micrometeorological measurements; and (b) differences over different areas of vegetation cover, from aircraft measurements (from André et al. 1988).

represented as weighted averages of the vegetation present. The forested portion, which was more organized, with length scales greater than 10 km, was observed to generate an ensuing response in the atmospheric boundary layer. The presence of this response, albeit captured through the mesoscale models, none the less suggests that convective processes and perhaps even a 'forest breeze' may intervene to confuse the synthesis of the area-average variables.

The First ISLSCP Field Experiment (FIFE)

The primary objective of FIFE (Sellers *et al.*, 1988b) is to provide (for the first time) data at a range of scales, simultaneously and at the same place, descending from the satellite scale, through airborne measurements, to near-surface micrometeorological measurements, and finally to basic process studies on plants and soils. The unique set of data provided in this way therefore includes satellite data from the NOAA, GOES, LANDSAT and SPOT satellites; TIMs and microwave data, collected with a C130 aeroplane, along with that from a range of helicopter-borne optical sensors, and airborne fluxes from three aircraft. The FIFE data also includes a uniquely comprehensive set of measurements of surface exchange made with micro-meteorological instrumentation at 20 individual points across the 15×15 km square study site. These are in turn complemented with detailed measurements of the optical properties of the prairie-grass vegetation, and regular measurements of surface and sub-surface soil moisture.

Early results from the micrometeorological network (Shuttleworth *et al.*, 1989; Smith *et al.*, 1992) show that the proportion of energy available at the ground which is used to support evaporation is fairly constant through daylight hours and, in consequence, the mid-day value of this proportion, called the evaporative fraction, is strongly correlated with the all-day average. Clearly this is a significant result in the context of remote sensing systems, since these normally provide only a snapshot of surface interactions. A second important result from the remote sensing standpoint, is that this evaporative fraction proved to be only weakly related to surface parameters such as leaf-area index, the proportion of green leaves present and the surface soil moisture, and to the remote sensing surrogates of these parameters. This finding is evidence that surface exchange processes are perhaps more uniform than remote sensing images might suggest (Shuttleworth, 1991a). The imposition of energy and mass conservation through the evaporation process, and the presence of compensating processes occur in practice, such as alternate evaporation from either vegetation or soil. The extreme variability of surface features apparent in remotely-sensed images may therefore provide a false representation of the true variability in surface exchange.

One very positive result from FIFE (Sellers *et al.*, 1988a) is the fact that the solar and photosynthetically active radiation measured by the surface micrometeorological network was capable of reasonable estimation using satellite data (Figure 9.6). Moreover, preliminary results suggest a worthwhile correlation between the net carbon dioxide uptake at those sites where measurements were made and the 'simple ratio' measured from the helicopter. However, the correlation between measured carbon dioxide uptake and water loss at these same sites shows less satisfactory agreement, perhaps again reflecting the influences of energy and mass conservation on the evaporation process.

The fact that there were so many surface measurements involved in the experiment provided insight into the reliability of micrometeorological instrumentation in general (Shuttleworth, 1991b). An important consequence was the recognition that the net radiometers used in the experiment were prone to systematic errors and non-linearity of the order of 5–10 per cent, thus indicating the need for better selection and calibration of such instruments. Further, the level of agreement between the different energy partition instrumentation suggests that the micrometeorological systems presently available can provide definition of energy partition to an accuracy no better than 10 per cent.

FIFE provided a warning against the use of radiometric surface temperature as a means of extrapolating point measurements of surface fluxes (Stewart *et al.*, 1989), the level of variability from site to site and its systematic error at individual sites being such as to throw doubt on the utility of this technique. Features of the temperature data can be explained given sufficient information on surface characteristics, but the additional data requirement is such that the technique itself is perhaps more a sink, rather than a source, of information. The experiment also revealed significant differences in surface energy exchange as latent and sensible heat, as measured by aircraft, in comparison with the surface measurements, suggesting a systematic under-estimation in the order of 40 per cent by the airborne systems (Sellers *et al.*, 1988b). The origin of this systematic error is not yet clear, but its presence in other experiments, such as HAPEX-MOBILHY and the LOTREX experiment, indicates that there is much to learn about the appropriate flight-lengths required in experiments of this type.

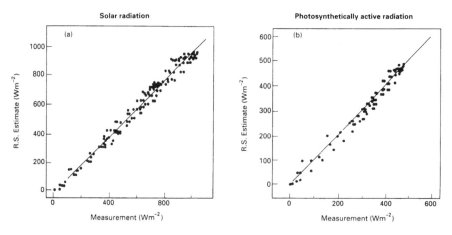

Figure 9.6   Data from FIFE illustrating good agreement between estimates of (a) solar radiation, and (b) photosynthetically active radiation derived from remote sensing data, and that measured by the micrometeorological surface network (from Sellers et al., *1988).*

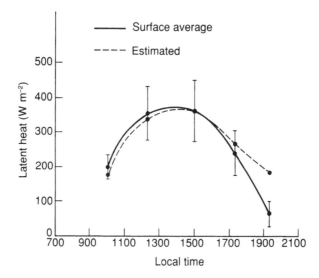

Figure 9.7   Area-average latent heat flux for the FIFE site derived indirectly, from energy conservation models of the atmospheric boundary layer, compared with that measured at the surface (from Munley, 1990).

One very positive result from FIFE is the fact that quite simple models of the boundary layer proved capable of representing the influence of surface controls on boundary layer development (Shuttleworth, 1991a). Inverting such models potentially provides a means for deducing area-average surface fluxes, and merits further investigation. Preliminary results indicate the value of using the energy balance of the boundary layer in this way (Figure 9.7).

General comments

Previous large-scale land/atmosphere studies have certainly been successful in as much as they have provided single-point calibration of tropical forests (from the ARME experiment), prairie grass (from the FIFE experiment), and mixed agricultural crops in central Europe (from the LOTREX experiment). The HAPEX-MOBILHY experiment provided calibration for characteristic southern European vegetation covers, and moreover provided insight into aggregation mechanisms for area-average surface-energy fluxes. It suggested that surface fluxes from disorganized vegetation cover could be estimated as a simple weighted average, while land cover organized at land scales greater than 10 km required synthesis of area-average fluxes via mesoscale modelling techniques.

Results from the FIFE experiment are not yet complete, but the experiment has provided a uniquely valuable source of data, and through this a test bed for the application of remote sensing methods in climate-related problems.

The future land surface programme

Future land surface observational studies will proceed through WCRP's *Global Energy and Water cycle EXperiment* (GEWEX) (WMO/ICSU, 1991) and IGBP's Core Project *Biological Aspects of the Hydrological Cycle* (BAHC) (IGBP, 1990). They will be essentially interdisciplinary, involving the scientific specialists involved in and propagating both these international programmes.

The Mesoscale Field Experiment Programme

The primary potential product of the Mesoscale Field Programme is the provision of Soil/Vegetation/Atmosphere Transfer Schemes (SVATS). These are mathematical representations, expressed as computer algorithms, which represent the energy, water, and perhaps in due course, bio-geochemical interaction of the interface between land surfaces and the atmosphere above. As outputs they provide the fluxes of energy and water into the atmosphere, and also route waters in the surface and sub-surface zone as runoff soil moisture and aquifer recharge. As input they require frequent values of near-surface meteorological variables, such as radiation components, temperature, humidity and precipitation, and they also require prescription of the parameters involved in the formulae. It is the provision of these parameters which is the central need.

In practice, models of global processes require representation of land/atmosphere interactions at very large area scales, and this is complicated by the fact that the terrestrial surfaces exhibit significant heterogeneity. Such heterogeneity arises through surface variation differences in vegetation, soils and topography, and is also generated by the atmosphere, through convective rainstorms and differing cloud cover. The representation is further complicated by the presence of advective processes, for instance advection of surface waters, but also very significant advection in the atmosphere. It is perhaps some consolation that atmospheric advection is usually in response to surface inhomogeneity, and generally acts to moderate the consequence of that inhomogeneity.

Area-average SVATS remain one-dimensional models, but the parameterization of these models is necessarily 'tuned' to provide the required area-average description and, in so doing, these parameters clearly can lose their local physical and physiological significance. The provision of these parameters is none the less a simple engineering exercise, given the availability of suitable data against which they can be optimized. The provision of such data is, however, a very significant problem.

Data requirements for calibration are simultaneous measurements of surface exchanges and controlling meteorological variables averaged over large areas, both sustained over long periods and frequently sampled at approximately hourly intervals. Such data do not exist, but a methodology has been evolved to provide them in surrogate form through mesoscale model synthesis.

The Joint IGBP/WCRP Working Group on Land Surface Experiments (WMO/ICSU, 1990) at its first meeting in Wallingford in 1990, specified the priority topics and preferred locations, and outlined an experimental format suitable to investigate provision of area-average SVATS. The experimental programme envisages two experiments investigating desertification in Spain and Niger, an experiment into tropical deforestation in Amazonia, and experiments designed to investigate the interaction of northern biomes, particularly the Boreal Forest in Canada, and Tundra at a location as yet unspecified (Figure 9.8). The relative timing of these experiments has been selected to allow the interested international community to participate in more than one (Figure 9.8).

The experimental format is illustrated in Figure 9.9. In these studies, typical 100 × 100 km areas will be defined, these areas being sufficiently large to include mesoscale meteorological activity. The variations in land cover and soil conditions over the large area will require careful definition because of their influence on exchange processes. Since the research issues are related to global change, site selection will be such that the area either includes existing eco-climatological gradients, as in Niger and Canada, or man made land-use change, as in Amazonia and Spain.

Subsidiary sites, which are catchments, typically with an area of 100 km², will be defined within this larger area at which long-term ecological and climate monitoring will occur for at least five (preferably ten) years, encompassing the year of major study. These areas also require careful selection to define and replicate variations in surface characteristics. The long-term effects of land-use change can be incorporated into such studies by space-for-time substitution. If the eco-climatological gradients demand it, as in the case of Boreal Forests, one or more of the ecological monitoring sites may have to be located outside the primary 100 × 100 km study area.

In addition to the eco-climatological monitoring at subsidiary sites, routine long-term data collection for the larger (100 × 100 km) study area will take place from a network of automatic weather stations, and from rain radar and satellite systems. Moreover, at least one, the most representative, subsidiary site will be chosen for long-term measurements of surface flux exchange and for catchment data collection.

During the main experimental year, data collection will increase dramatically. Routine data collection for the full study site will then include boundary layer sounding, with radiosondes and ground-based stations. Surface flux measurements will be made at each of the subsidiary sites for the characteristic classes of surface cover present there, and the hydrological data collection extended to cover all such subsidiary areas. For selected periods of the year these subsidiary sites will be overflown by aircraft carrying remote sensing instruments, and to provide airborne flux measurements. Experimental activity in the precursor experiment will be broadly of the same nature and intensity, but is likely to be restricted to just one of the subsidiary sites. Ecological studies of three types are required:

(1) definition of the spatial variation in vegetation and soil characteristics which are relevant to energy, water and trace gas flux and which can be remotely sensed over extensive areas,
(2) measurement of temporal variation (daily, seasonal, successional, cyclical, episodic, stochastic) in vegetation and soil characteristics affecting land–atmosphere interactions, and
(3) experimental studies in both field and laboratory to quantify the relationships of key vegetation and soil characteristics to variation in climate variables (CO_2, temperature, moisture, radiation) both singly and in combination.

Priority Topic	Location (Experiment)	Provisional Timing							
		1991	1992	1993	1994	1995	1996	1997
Desertification	Spain (EFEDA)	----	....	....	++++	....	....	
	Niger (HAPEX-Sahel)	....	++++	....	....	....	....	....
Boreal Forest Interaction	Canada (BOREAS)			----	++++	....	....	....
Tropical Forest	Brazil (?)	....	....	----	....	....	++++	....
Tundra Interaction	*(to be defined)*				*(to be defined)*			

Key: ---- Small-scale (single site) experiment
      .... Monitoring effort
      ++++ Large-scale (multiple site) experiment

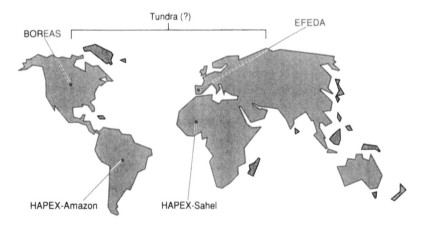

Figure 9.8   The priority topics, proposed locations and provisional timing of the Mesoscale Field Experiment Programme proposed by WCRP-GEWEX and IGBP-BAHC; and the proposed locations of the 100 × 100 km study areas on the globe.

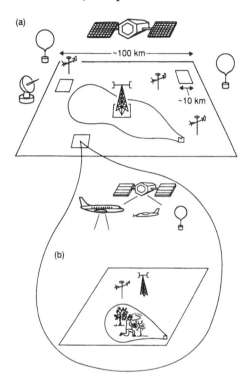

Figure 9.9 Experimental format for BAHC Mesoscale Field Experiments (a) for a site typically 100 × 100 km, long-term (5–10 year) monitoring using a mesoscale meteorological network, rain radar and satellite systems; for several subsidiary sites, long-term ecological monitoring; and for at least one of these subsidiary sites, preferably the most representative in terms of surface cover, long-term monitoring of surface fluxes and climate, hydrological catchment data and vegetation characteristics; (b) for several (2 or more) subsites nested in the larger area, monitoring lasting up to a year, perhaps in short, very intensive sessions, of near-surface fluxes and climate, canopy and soil processes and atmospheric profiles, with airborne measurements of boundary layer fluxes and intensive satellite and airborne remote sensing.

The modelling methodology underlying the Mesoscale Field Experiment Programme is illustrated in Figure 9.10. Its aim is to provide simultaneous data in the form of surface flux measurements for characteristic vegetation covers, boundary layer soundings and surface weather measurements to allow the use of coupled hydrological/atmospheric models. These will operate at a mesoscale, for example with a 10 km grid net, and their adequate performance will be further checked by the aircraft and remote sensing data, and the hydrological catchment data. Such models will be used in the development of parameterizations to allow for heterogeneities in the atmospheric variables (e.g. precipitation), and in the surface characteristics. The long time-series data provided by the surface meteorological network covering the site, by the rain radar system (and associated ground calibration), and by meteorological data from the surrounding area will provide the boundary conditions for the mesoscale model syntheses. The surface representations for the several different land cover types will be calibrated against the micrometeorological flux measurements and associated catchment data, while the coupled hydrological/atmospheric model will itself be validated against the atmospheric sounding and aircraft flux data, and against the catchment data.

Once calibrated and validated in this way, the mesoscale model can then be used to provide synthetic macroscale data as area–average surface fluxes, area–average near-surface weather variables, and surface and sub-surface runoff. These data can then be used to calibrate grid-scale representations. The inclusion of sampled eco–climatological gradients, or man-made changes within the study area will allow the synthesis of equivalent grid-scale parameterizations for a whole range of hypothetical mixes and spatial distributions of surface vegetation cover. The modelling methodology and the typical calibration for representative cover types can, moreover, be applied to provide simulation of changes elsewhere in the sampled biome, the relative mix of vegetation cover being then determined from remote sensing data.

A second important modelling focus in these studies will be to investigate also the reliability of the surface representations applied at the 10 km scale, using finer scale models operating with a grid mesh of 1 km or less. Such models will be used to evaluate the significance or otherwise of atmospheric

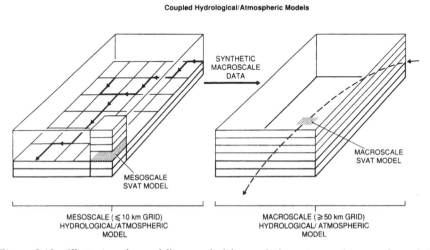

Figure 9.10   Illustrating the modelling methodology which envisages the use of coupled hydrological/atmospheric models operating at mesoscale to synthesize equivalent area-average macroscale parameters at macro (i.e. grid) scale from which to calibrate the effective values of parameters used in very large-scale area-average SVAT models.

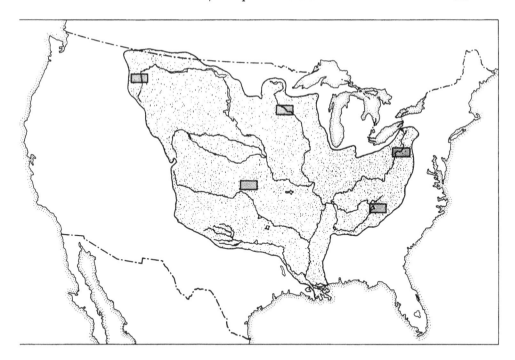

Figure 9.11 Schematic illustration of the GEWEX Continental-scale International Project (GCIP) to occur in the Mississippi basin, with focused attention on subsidiary areas (location as yet unspecified) to enhance the global relevance of the study.

advection as air moves from one plot of vegetation cover to the next, so refining the first stage of the 'scaling up' process from the field scale.

Continental-scale studies

The initiative for continental-scale activity currently lies within WCRP's GEWEX programme, and the science plan for a continental-scale study of the Mississippi basin is under active preparation, to be called the GEWEX Continental-scale International Project (GCIP). It is illustrated in Figure 9.11. The study will be jointly supported by major US agencies, but invites international participation, particularly in data interpretation and modelling aspects.

The primary objectives of GCIP, as so far defined (WMO/ICSU, 1991) are:

(1)   to develop and validate macroscale hydrological models and coupled hydrological/atmospheric models, to obtain a quantitative understanding of the energy and water cycles over extended land uses;
(2)   to develop and validate information retrieval systems, incorporating existing and future satellite observations with *in situ* measurements, to enable expanding GEWEX models and database to the global domain;

(3)   to provide a capability to translate the effects of a future climate change into impact on water resources on a regional basis.

The data potentially available through the GCIP is massive. In addition to the long-established and ongoing hydrological data networks throughout the Mississippi basin, and the extensive network of meteorological stations, there are plans for a nationwide network of surface-based weather radar systems, the National Weather Service 'NEXRAD' system, which are considered to be crucial to the project. The Mississippi is, of course, also the best mapped, digitally defined large river basin in the world, and heavily monitored by both aircraft- and satellite-borne systems. Moreover, the already substantial boundary sounding network could be easily extended to allow the large-scale determination of the hydro-meteorological interaction through determination of change in atmospheric energy and water balance. A very important part of the design of this experiment will be the creation of an appropriate data handling system to assimilate and distribute this vast data resource.

In the experimental design of GCIP, attention is currently being given to specifying and selecting subsidiary areas across the river basin on which more detailed monitoring and modelling can be focused. This is with a view to better establishing the global relevance of the study, and to allow the global-scale hydrological prediction skills developed in the Mississippi to be applied elsewhere.

Concluding remarks

The upcoming international programme overviewed in the preceding section has evolved over several years as the consensus plan of many international scientists. It has well-defined experimental formats, modelling methodology, objectives and priorities. The terrestrial community is steeled, motivated, committed and ready to go.

In the United Kingdom there are priming funds to engage in this programme under NERC's Terrestrial Initiative into Global Environmental Research (TIGER) programme, but there is now a need for funding commensurate with the complexity, importance and, not least, relevance of this research. The funding levels required are broadly comparable to those currently being deployed in other important areas of scientific endeavour such as nuclear physics, space science or astronomy. This research is no longer limited by ideas and plans, only by resources. To paraphrase Sir Winston Churchill, 'Give us resources; we can do the job'.

References

Andre, J.-C., Goutorbe, J.-P., Perrier, A., Becker, F., Besselmoulin, P., Bougeault, P., Brutsaert, Y., Carlson, T., Cuenca, R., Gash, J.H.C., Gelpe, J., Hildebrand, P.,

Lagouard, P., Lloyd, C.R., Mahrt, L., Mascart, P., Mazaudier, C., Noilhan, J., Ottle, C., Payen, M., Phulpin, T., Stull, R., Shuttleworth, W.J., Schmugge, T., Taconet, O., Tarrieu, C., Thepenier, R.M., Valencongne, C., Vidal-Madjar, D. and Weill, A., 1988, HAPEX-MOBILHY: first results from the special observing period, *Annals of Geophysics*, **6** (5), 477–92.

Charney, J.G., Quirk, W.J., Chow, S.H. and Kornfield, J., 1977, A comparative study of the effects of albedo change on drought in semi-arid regions, *Journal of Atmospheric Science*, 1366–85.

Dickinson, R.E., 1989a, Implications of tropical deforestation for climate: a comparison of model and observational descriptions of surface energy and hydrological balance, *Philosophical Transactions of the Royal Society of London*, **B234**, 423–31.

Dickinson, R.E., 1989b, Modelling the effects of Amazonian deforestation on regional climate: a review, *Agricultural and Forest Meteorology*, **47**, 339–47.

IGBP: The International Geosphere-Biosphere Programme, 1990, *A Study of Global Change. The Initial Core Projects*. IGBP, Stockholm, IGBP Report No. 12.

Mascart, P., Taconet, O., Pinty, J.-P. and Mehrez, M.B., 1990, Canopy resistance formulation and its effect in mesoscale models: a HAPEX perspective, *Agricultural and Forest Meteorology*, **54**, 319–35.

Munley, W.G. and Hipps, L.E., 1990, Estimation of regional evaporation for a tall grass prairie from measurements of properties of the atmospheric boundary layer. *Water Resources Research*, **27**, 225–230.

Pinty, J.-P., Mascart, P., Richard, E. and Rosset, R., 1989, An investigation of mesoscale flows induced by vegetation inhomogeneities using an evaporation model calibrated against HAPEX-MOBILHY data, *Journal of Applied Meteorology*, **28**, 976–92.

Sato, N., Sellers, P.J., Randall, D.A., Schneider, E.K., Shukla, J., Kinter III, J.L., Hou, Y.-T. and Albertazzi, E., 1989, Effects of implementing the simple biosphere model in a general circulation model, *Journal of Atmospheric Science*, **46** (18), 2757–82.

Sellers, P.J., Hall, F.G., Strebel, D.E., Kelly, R.D., Verma, S.B., Markham, B.L., Blad, B.L., Schimel, D.S., Wang, J.R. and Kanemasu, E., 1988a, *First ISLSCP Field Experiment, April 1988, Workshop Report*, Code 624, NASA/Goddard Space Flight Center, Greenbelt, MD 20771, USA.

Sellers, P., Hall, F.G., Asraf, G., Strebel, D.E. and Murphy, R.E., 1988b, The First ISLSCP Field Experiment (FIFE), *Bulletin of the American Meteorological Society*, **69**, 22–7.

Sellers, P.J., Shuttleworth, W.J., Dorman, J.L., Dalcher, A. and Roberts, J.M., 1989, Calibrating the simple biosphere model for Amazonian tropical forest using field and remote sensing data. Part 1: average calibration with field and remote sensing data, *Journal of Applied Meteorology*, **28**, 727–59.

Shukla, J. and Mintz, Y., 1982, Influence of land-surface evapotranspiration on the Earth's climate, *Science*, **215**, 1498–1501.

Shukla, J., Nobre, J.C. and Sellers, P.J., 1990, Amazon deforestation and climate change, *Science*, **247**, 1322–25.

Shuttleworth, W.J., 1988a, Evaporation from Amazonian rain forest, *Philosophical Transactions of the Royal Society of London*, **B**, 233–346.

Shuttleworth, W.J., 1988b, Macrohydrology—the new challenge for process hydrology, *Journal of Hydrology*, **100**, 31–56.

Shuttleworth, W.J., 1991a, Insight from large-scale observational studies of land/atmosphere interactions, in Wood, E.F. (Ed.), *Land Surface Atmosphere Interactions: Parameterization and Analysis for Climate Modelling*, New York: D. Reidel.

Shuttleworth, W.J., 1991b, The modellion concept, *Reviews of Geophysics*, **29** (4), 585–606.

Shuttleworth, W.J., 1991c, *The Role of Hydrology in Global Science*, International Association of Hydrological Sciences Publication No. 204, Wallingford, Oxfordshire, UK, 361–75.

Shuttleworth, W.J. and Dickinson, R.E., 1989, Comments on 'Modelling Tropical Deforestation: A Study of GCM Land-Surface Parameterizations' by R.E. Dickinson and A. Henderson-Sellers, *Quarterly Journal of the Royal Meteorological Society*, **115**, 1177–9.

Shuttleworth, W.J., Gash, J.H.C., Lloyd, C.R., Moore, C.J., Roberts, J.M., Marques, A. de O., Fisch, G., Silva, V. de P., Ribeiro, M.N.G., Molion, L.C.B., de Sa, L.S.A., Nobre, J.C., Cabral, O.M., Patel, S.R. and Moraes, J.C., 1984, Eddy correlation measurements of energy partition for Amazonian forest, *Quarterly Journal of the Royal Meteorological Society*, **111**, 1143–62.

Shuttleworth, W.J., Gurney, R.J., Hsu, A.Y. and Ormsby, J.P., 1989, *FIFE: The Variation in Energy Partition at Surface Flux Sites*, International Association of Hydrological Science Publication No. 186, Institute of Hydrology, Wallingford, Oxfordshire, UK, 67–74.

Smith, E.A., Hsu, A.Y., Crosson, W.L., Field, R., Fritschen, L.J., Gurney, R.J., Kanemasu, E.T., Kustas, W., Nie, D., Shuttleworth, W.J., Stewart, J.B., Verma, S.B., Weaver, H. and Wesley, M., 1992, Area-average surface fluxes and their time-space variability over the FIFE experimental domain, *Journal of Geophysical Research*, (in press).

Stewart, J.B., Shuttleworth, W.J., Blyth K. and Lloyd, C.R., 1989, FIFE: A Comparison Between Aerodynamic Surface Temperature and Radiometric Surface Temperature Over Sparse Prairie Grass. *Proceedings of the 19th Conference on Agriculture and Forest Meteorology and 9th Conference on Biometeorology and Aerobiology*, 7–10 March 1989, Charleston, South Carolina: American Meteorological Society, 144–6.

Sud, Y.C., Shukla, J. and Mintz, Y., 1985, *Influence of Land-Surface Roughness on Atmospheric Circulation and Rainfall: A Sensitivity Experiment with a GCM*, NASA Technical Memo. 86219, NASA/Goddard Space Flight Center, Greenbelt, Maryland, USA.

WMO/ICSU, 1990, Report on the First Meeting of the Joint IGBP/WCRP Working Group on Land-Surface Processes, WMO/TD-370, World Meteorological Organisation, Geneva, Switzerland.

WMO/ICSU, 1991, Scientific Plan for the Global Energy and Water Cycle Experiment, WMO/TD-376, World Meteorological Organisation, Geneva, Switzerland.

Chapter 10

Monitoring global tropical deforestation: a challenge for remote sensing

J.P. MALINGREAU, M.M. VERSTRAETE
and F. ACHARD

Institute for Remote Sensing Applications,
Joint Research Centre of the CEC,
Ispra, Italy

Global tropical deforestation—the issues

Large-scale transformations of the tropical forest biome are often presented as one of the major ecological issues of the final decade of the twentieth century. Wanton and irreversible deforestation for logging purposes, slow encroachment of agriculture into forest lands, and unexpected and catastrophic damage from uncontrolled burning, have all contributed to focus the attention of the public at large on the fate of these unique tropical forest ecosystems. In the last decade the scale of awareness, interest and concern has become global. The range of possible impacts of forest removal has been given ample publicity, and environmental consequences have been examined with varying degrees of confidence at scales ranging from local to regional and even global. Predictions related to habitat destruction, reduction in species diversity, loss of soil and water resources, changes in local and regional climates, and impacts on local population, currently range from the benign to the catastrophic. It is, indeed, unusual to find statements on world forest resources and deforestation that are entirely objective and free of philosophical underpinnings. The fact that, in an increasingly ecologically-conscious world, the forest is seen as the symbol of environmental purity does not facilitate an objective evaluation of the facts.

In preserving the tropical forests of the world, a parallel can be drawn with global warming issues: 'Let us save what can still be saved because we have too much to lose', goes one group of proponents. 'Let us keep watching and be sure of our facts before we deprive needy countries of a direct and precious renewable resource', say others. In between, schemes are designed to combine the best of two worlds and exploit in a 'sustainable' way what often represents the sole means of survival for large and deprived populations. This middle way reflects in a sense man's ambivalent historical relationship with the forest. This is not the place to review all the arguments or cite examples in favour of one or another position. Yet, the above can

121

serve to set the stage for a definition of needs with respect to global forest resource information. The stage is crowded, indeed. Requirements for facts and figures are as varied as opinions and positions of the issues—they are as complex as the forests resources themselves, which are 'bewildering in their extent and complexity' (Mather, 1990).

While the reality of a pervasive and accelerating tropical deforestation is not in dispute (FAO, 1988), most statements related to this phenomenon as a global issue are based on estimates which are difficult to verify at such a scale and are indeed open to question (WRI, 1990). World statistics on forests and deforestation are attained through the process of aggregating national or local statistics, all carrying their inherent weaknesses and inaccuracies. There is a plethora of information on forest resources but they lack per force uniformity of coverage, quality and content; such information is scattered and diverse, compilation is difficult and synthesis all too rare. A first set of questions can be raised in relation to the definition of one or a few forest parameter(s) which could represent a unifying concept in forest assessment, the design and operation of a monitoring system which could systematically collect and update such information at various scales, and the feasibility of synthesizing the derived information to help support a sensible conservation plan.

Turning now to space data collection platforms, we have to ask ourselves if these much-vaunted global sensors of today and tomorrow can support an information system capable of answering the spectrum of needs outlined above. To analyse this question we review the kind of information which can be obtained from current remote sensing systems with respect to the forest canopy, and then proceed to describe the characteristics of a possible global forest monitoring set-up and how it could be operated. We also emphasize the importance of modelling at every step in the analysis of remote sensing data.

Forest parameters and remote sensing systems

At regional to global scales, a list of surface parameters to be included in a monitoring system is not easily compiled. Indeed, most information requirements are naturally expressed at more 'human' scales, that is those which correspond most clearly to visual observation or measurement on the ground. Forest classification schemes reflect such a perspective, as they are based upon criteria such as floristic composition and structure. When expanding such point observations, logic dictates that one looks, by statistical sampling or otherwise, for the collection of points which satisfies the given set of characteristics. Vegetation mapping has classically proceeded on such a principle. Similarly, from the earliest days of air photography, remote sensing analysis has been concerned with the search for and identification on images of vegetation types previously characterized in classic surveys. The search for specific and unique spectral signatures has been a building-block of the approach.

At scales ranging from regional and continental to global surveys, there is now a need for a new classification scheme simultaneously compatible with the level of information required and the capabilities of space-borne sensors. Two lines of investigation can be contemplated. Firstly, one might envisage an approach involving selection of the appropriate hierarchical level of classification, with analysis conducted at generalized levels. Such is the case in recent forest assessments (Myers, 1989; WRI, 1990) and of proposals to carry out a forest:non-forest classification of land cover at global levels (TREES, 1990). There is undoubtedly a huge demand for a global data set interpreted in those binary terms; it does not yet exist, but would be highly valuable as a baseline for an overall assessment of the deforestation processes. Despite the apparent simplicity of the approach many conceptual difficulties remain, among others of semantics. The very concepts which are at the core of the debate, such as 'deforestation' and 'forest degradation', lack universal definition. When is a tropical forest canopy closed or open? When is it pristine or degraded? Obviously, the term 'deforestation' means different things to different people. Selective logging is a form of sustainable management which leaves the forest cover in a state of potential regeneration, yet for the committed conservationist this activity will constitute 'deforestation'. A climate modeller will not be concerned about the type of forest, but will be interested in albedo change, roughness of the canopy, radiation balance and other physical parameters. The ecologist's interest will range from species inventory to net primary productivity of the ecosystem. A common denominator must be found if the variety of users are to be satisfied at least at the more generalized level of interest. Starting with the classical inventory of needs does not seem to be a promising way to define such a common denominator and, maybe, there is room here for new thinking on the role of remote sensing data as a unifying driver.

A possibly novel approach to forest classification relates to a more intensive use of data derived from space sensors. At the outset, it is important to remember that such data are simply of a radiative nature only; any vegetation-related information which can be extracted from the set is always expressed first as a radiative property of the canopy. A corollary is that any biospheric element or process which does not induce a measurable change in the radiative characteristics of the canopy cannot be detected. It follows that space observations must be interpreted and analysed with models which can, in a systematic and quantitative fashion, trace and simulate the links between a set of radiometric measurements and the state of a plant community. A range of models of varying degrees of sophistication can be identified. Empirical models attempt to establish direct links between remotely-sensed data and ground observations or proxy for the desired information (for example, relationships between vegetation indices and green biomass via statistical regression techniques). Such an approach has proved useful for target identification, mapping, event detection or exploratory investigations. Physical models link the sensor signal to a surface parameter. Biospheric models attempt to relate a specific surface parameter to a plant process (or its by-product), as in attempts to link net primary productivity, vegetation indices and the absorption of photosynthetically-active radiation. At this stage, physical models are still under development and cannot be applied in an operational context to any kind of vegetation. From our point of view, if we wish to extract new information from satellite data sets, we must ask how radiation data can be exploited not only to identify objects but also to provide continuous physical measurements of vegetation without explicit reference to a pre-set classification. Two examples illustrate this point:

(1)   it has been shown that spectral contrasts, whether related to albedo or to brightness temperature, can lead to a spatial separation between the evergreen tropical forest cover and the surrounding agricultural land. Within the forest canopy, degradation will translate into a continuum of radiometric changes which can only be identified as fields of increased albedo, increased surface temperature and the like. In the first instance a reference to a binary forest—non-forest classification is sufficient; in the second, thresholds of forest degradation must be decided upon since the radiometric information is non-discrete;

(2)   forest seasonality is a character explicitly used in tropical forest classification (i.e. tropical deciduous forest, semi-deciduous forests, dry seasonal forests, monsoon forest, etc.). While such classification is binary in the sense that it considers that a forest is seasonal or not, it avoids the question of the intensity or even facultativeness of seasonality processes (Woodward, 1987). Time series of satellite-derived vegetation index data are now available over ten years; they can be used as an indication of the seasonality of the photosynthetic process (Justice *et al.*, 1985; Tucker *et al.*, 1985). Again, the data provided by the satellite show a continuum of signal amplitudes from the 'evergreen' to the fully seasonal pattern. Such forest canopy related information is highly relevant for a series of analyses (Malingreau, 1990); seasonal flushes affect albedo and evapotranspiration. Interannual comparison indicates the impact of climate on the tropical forest vegetation. Drought and stress related to seasonal effects are also an indication of forest susceptibility to fire, a common agent of land clearing in the tropics.

These observations point to the need for a new look at remote sensing data for tropical forest monitoring. This new perspective should combine the following features:

(1)   measurement of spectral characteristics in wavelength bands ranging from the visible to the thermal infrared and beyond to the microwave region. Modelling of the signal in order to extract physical surface properties;

(2)   development of synergy between various channels (that is, the combined used of different channels must yield higher information return than the sum of individual analyses);

(3)   use of low resolution data sets to provide repetitive and global coverage.

Currently, the use of AVHRR data adhering to the principles set out in the following section, can lead to the extraction of information on tropical forest cover. It is worth noting that, to be innovative, such a data acquisition scheme must be seen in the perspective of a continuous and global exercise and not, as in the past, for a few selected applications.

A global tropical forest monitoring system

The need for updating baseline information on tropical forests of the world has been a major drive in the FAO/UNEP Tropical Forest Resources Assessment Project initiated in 1979. The exercise led to the production of the widely-known Forest Resources Assessment (Lanly, 1982). The updated version of the project (Singh, 1990) is intended to produce reliable estimates of forest cover in the tropical zone for the year 1990, and of the rates of change in forest cover area during the period 1981–90. The work is based on a survey design which includes existing reliable data from archives and from satellite remote sensing. Ultimately, a continuous forest monitoring system is to be implemented.

Other attempts have been made to produce up-to-date assessments of the world tropical forest cover (Myers, 1989; WRI, 1990; German Bundestag, 1990). However, their analyses fall beyond the scope of this paper. The initiative of the World Conservation Union (Collins, 1990) to compile existing forest maps and harmonize the information contained therein to produce an atlas, is noteworthy. A Geographical Information System is used to integrate various sources of information with respect to topography, nature reserves, ecosystem boundaries, etc. The result is comprehensive and can be considered to represent the current state of knowledge.

The following subsections describe a remote sensing observation system for tropical forest monitoring essentially based upon AVHRR data from the NOAA satellite series. The advantages of that particular sensor for global vegetation monitoring have amply been expounded elsewhere (Justice *et al.*, 1985) and the technology need not be described here. The monitoring scheme which is proposed here has been progressively refined, from the first AVHRR experiments over tropical forest areas, through a continuous process of consultation within the framework of the UNEP *ad hoc* Group of Forestry Experts and the International Space Year–World Forest Watch Project.

The three components of the proposed activity (Figure 10.1) are:

(1)   a baseline inventory of tropical forest cover using a comprehensive collection of AVHRR data for the year 1989-91;
(2)   a monitoring system for change detection; and
(3)   a series of models for studying tropical deforestation dynamics and impacts.

Baseline inventory

Recent experience has shown that low resolution AVHRR data can, when properly selected and processed, provide substantial amounts of information on the distribution of the closed forest cover of the tropics (Tucker *et al.*, 1985; Malingreau and Tucker, 1988; Malingreau *et al.*, 1989; Nelson and Holben, 1986; Achard and Blasco, 1990).

The objective is now to use the existing experience to carry out the analysis over the full forest area of the tropics. The following are guidelines on procedure (TREES, 1991):

(1) secure the so-called wall to wall coverage of the tropical belt by recent (post-1989) AVHRR data. Currently, this can only be done using reception facilities dispersed around the world. The lack of a systematic procedure for collecting such data is at the source of logistic problems which do not render the operation straightforward;

(2) pre-process the appropriate AVHRR data sets according to acceptable standards which relate to radiometric calibration and geometric rectification (Teillet, 1990);

(3) analyse the processed data set in order to delineate areas of remaining tropical moist forest formations and assess their degree of disturbance. The vegetation classes to be retained in the analysis have been listed in Figure 10.2;

(4) validate of the analysis using high resolution satellite data sets (LANDSAT, SPOT) and ancillary field reports and visits;

(5) import the final information into a GIS which contains levels of basic land-related information such as topography, river network, and major access roads. This GIS can be filled with additional information as the need arises for more specific or detailed analysis.

While the implementation of the steps above can be considered as operational, the inclusion of seasonal information derived from time series of Global Area Coverage (GAC) data requires further developments. These are considered important because, as already mentioned, they will add a new dimension to the analysis of the tropical vegetation canopy. The articulation of these various satellite products is shown in Figure 10.3.

The new perspectives for tropical forest inventories opened by microwave

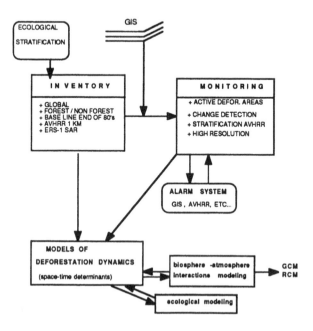

Figure 10.1   Inventory, monitoring and modelling system.

instruments are to be examined within the framework briefly described above. We have yet to ascertain, through pilot studies using ERS-1 products, how the information derived from microwave instruments will be integrated with results of analyses of optical-infrared data. Two main avenues of investigation seem possible: merging of the two sets of information and

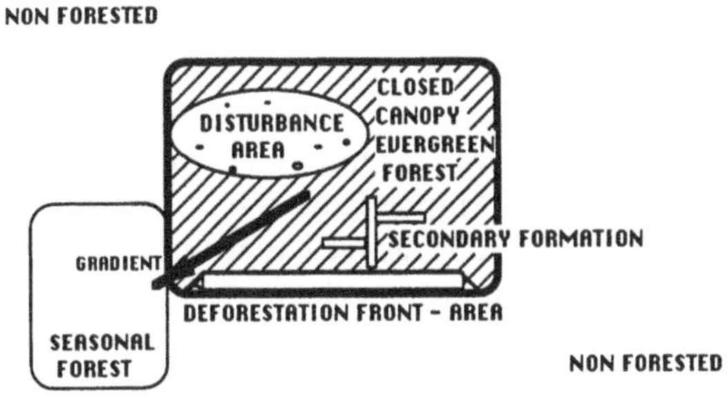

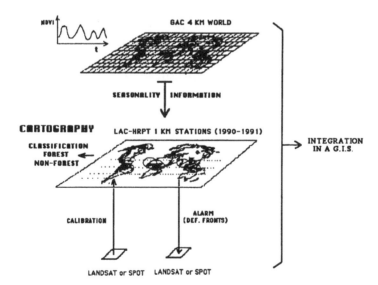

Figure 10.2   Components of forest cover classification.
- *closed canopy rainforest—evergreen (> 90 per cent cover)*
- *disturbed forest where regional forest cover is between 60 per cent and 90 per cent (various patterns)*
- *Deforestation front or area where forest cover comprises between 10 per cent and 60 per cent*
- *Non-forest area where forest cover <10 per cent*
- *Closed secondary regrowth*
- *Seasonal forest (gradients)*

Figure 10.3   AVHRR-GAC, AVHRR-LAC and LANDSAT/SPOT image data relationships.

addition of a third derived from their synergistic analysis, or comparison and adjustment of results using one set for validation of the other. Both approaches require further development.

Monitoring of active deforestation

Critical information on tropical forests is not related to their distribution but to changes occurring in their distribution. Monitoring of forest areas must, therefore, be oriented toward the detection, identification and measurement of transformations taking place in the areal extent of forest cover. The discrepancy between the global perspective of inventory and the often local nature of changes, must be emphasized here. There is thus a need for a technology which must be global as well as attuned to indicators of change, and precise enough to yield figures on changes. There is no single instrument which can deliver such order and a package has to be developed.

Currently, the approach to global forest monitoring can be envisaged as follows:

(1)   once the basic inventory of forested surfaces is available, the most active deforestation areas are identified using multisource information among which is the NOAA–AVHRR data, regularly collected by the local stations. A series of satellite-derived indicators of deforestation activities is currently under analysis (Malingreau, 1990); it include fires, opening of roads, degradation of the seasonal vegetation index signal, etc.;

(2)   high resolution data derived from LANDSAT and SPOT satellites are then used to measure rates of transformation of the canopy and to interpret such measurements in terms of 'deforestation'. At this stage, change detection techniques must be called upon to standardize and streamline the analysis.

The above procedure will lead to the elaboration by iteration of a world 'active deforestation' map superimposed upon the reference forest inventory.

Understanding remote sensing measurements in terms of ecosystem processes

Inventory and monitoring activities must not become ends in themselves, but should lead to a better understanding of tropical forest ecosystems and their role in the global environment. Tropical forest ecosystems are notoriously complex in their structure, species composition and functioning. The contribution of remote sensing to their study will touch only a minute fraction of the parameters needed for progressing in such understanding. Yet, observations from space provide new perspectives on Earth surfaces with respect to geographical scales, frequency of data acquisition and types of data. Comparative advantages of such kind must therefore be exploited if new insights are to be achieved. We see three main areas of investigation opening up in this respect. The following must not be seen as an exhaustive review of these issues but rather as a preliminary agenda for research.

(1)   Spatio-temporal models of deforestation. Deforestation patterns can be considered

as representing the combined response of the forest system to a series of pressures and constraints related to human activities. For example, the opening of the Amazon Basin in the mid-1980s can be principally associated with socio-economic conditions in Brazil (search for agricultural land or ranch opening for tax abatement purposes), and the pattern of deforestation is closely associated with the opening and even paving of access roads (Fearnside, 1986) and soil suitability for agricultural activities. The availability of satellite data allows the monitoring of these trends over large areas; the inclusion in a Geographical Information System of the major factors of human activity in the forest can lead to the evaluation of future spatio-temporal trends under a series of socio-economic scenarios. Remote sensing data can serve as a base to initiate such a model and verify its predictions. Each region of the tropics will present a different set of controlling conditions and driving variables. The global monitoring of vegetation thus requires, among other things, that the flow of information between the local, detailed observation level 'and the satellite, coarse resolution level must be adequately understood. The information content of sensor pixels of differing sizes must, therefore, be carefully assessed so as to select the most appropriate resolution for the detection and monitoring of local deforestation features (Malingreau and Bellward, 1991).

(2)   Tropical forest cover and climate models. An agenda for investigations on this topic can rapidly become very vast and again, the objective is to assess the extent to which satellite data can provide relevant measurements. Foremost, for continental studies, is the derivation, through inversion of spectral signals, of canopy characteristics influencing biosphere–atmosphere interactions, such as albedo and its seasonal variation, canopy roughness and skin temperature (Verstraete and Pinty, 1991). Spatial gradients in sensible heat flux resulting from deforestation-related contrasts in vegetation cover, appear especially critical in the formation of thermally-forced wind circulation (Pielke *et al.*, 1991). The relative distribution and size of openings in the forest canopy, and their orientation with respect to rain-bearing air masses (Salati amd Vose, 1984) must also be explored. Another domain of application relates to the analysis of time-series of measurements (albedo, vegetation index, surface temperature, etc.) for deriving information on the seasonality of the forest cover and its response to climatic determinants. Because of the likelihood that man can and is likely to continue to transform large tracts of tropical forest, assessing the influence of mesoscale land cover variations on weather and climate has acquired some urgency.

(3)   Forest ecological models. Habitat fragmentation or gap distributions are intrinsic characteristics of the forest ecosystem. Such spatial properties lend themselves to observation from space, although limited ground resolution and spectral contrasts may preclude the collection of relevant data on significant gap size classes in the canopy. Remote sensing techniques can help in defining the topology of forest–non-forest boundaries. Attempts to map primary productivity using the norma-lized difference vegetation index have been made using statistical regression models (Goward *et al.*, 1985). Process-based models using a conservation of energy approach have also been investigated. Time series of satellite data related to canopy characteristics can thus form the basis for spatial modelling of vegetation response to climate scenarios. Interpretation of forest seasonality in terms of susceptibility to degradation can also be attempted using the same data set (Malingreau, 1990).

Conclusions

Observations from space increasingly provide new vistas on the tropical forests of the world. They can support a wide range of investigations relating to the distribution, change therein and functioning of forest ecosystems. It must be recognized, however, that remote sensing technology cannot provide all the necessary data, and its limitations have to be carefully assessed. Remote sensing approaches will never directly provide information on cover characteristics or processes which do not influence the radiative properties of the surface (Verstraete and Pinty, this volume, Chapter 16). Furthermore, the technology and expertise available today for data analysis does not give a free choice in the necessary compromise between ground resolution, frequency of acquisition and global nature of the coverage. Spatial resolution and frequency of coverage are still logistically competitive and tend even to be exclusive of one another. A challenge for remote sensing research is to assess whether global information remains pertinent at higher resolutions. Finally, if one can consider that inversion problems related to the remote measurement of physical surface parameters will one day be solved, this achievement will represent the start of an investigation into biological processes which represents the ultimate purpose of the effort. From empirical analysis of long-term and global satellite data sets, to the development and operation of process-based models, the task of better exploitation of satellite data is challenging indeed. The case of the tropical forest is particularly critical since, through an enhanced understanding of its nature and vulnerability, we could potentially help in the preservation of this unique and most precious biome.

References

Achard, F. and Blasco, F., 1990, Analysis of vegetation seasonal evolution and mapping of forest cover in West Africa with the use of NOAA AVHRR HRPT data, *Photogrammetric Engineering and Remote Sensing*, **56**, 1359-65.

Collins, M. (Ed.), 1990, *The Last Rain Forests*, London: Mitchell Beazley Publishers, 200 pp.

Cross, A.M, Settle, J.A., Drake, N.A and Paivinen, R.T.M., 1991, Subpixel measurement of tropical forest cover using AVHRR data, *International Journal of Remote Sensing*, **12**, 1119-29.

FAO, 1988, *An Interim Report on the State of Forest Resources in Developing Countries*, Rome: Forest Resources Division, Forestry Department.

Fearnside, P.M., 1986, Spatial concentration of deforestation in the Amazon Basin, *Ambio*, **15**, 74-81.

German Bundestag, 1990, *Protecting the Tropical Forests. A High Priority International Task*, Second Report of the Enquete Commission, 'Preventive Measures to Protect the Earth's Atmosphere', of the 11th German Bundestag, Bonn, 968 pp.

Goward, S.N., Dye, D. and Tucker, C.J., 1985, North American vegetation patterns observed with Nimbus-7 Advanced Very High Resolution Radiometer, *Vegetatio*, **64**, 3-14.

Justice, C.J., Townshend, J.R.G., Holben, B.N. and Tucker, C.J., 1985, Analysis of the phenology of global vegetation using meteorological satellite data, *International Journal of Remote Sensing*, **6**, 1271-1318.

Lanly, J.P., 1982, Tropical forest resources, *FAO Forestry Paper*, **30**, Rome: FAO-UNEP, 106 pp.

Malingreau, J.P., 1990, The contribution of remote sensing to the global monitoring of fires in tropical and subtropical ecosystems, in Goldammer, J.G. (Ed.) *Fire in the Tropical Biota. Ecological Studies,* **84**. Amsterdam: Springer Verlag, 337-70.

Malingreau J.P. and Belward, A.S., 1992, Scale considerations in vegetation monitoring using AVHRR data, *International Journal of Remote Sensing,* in press.

Mather, A.S., 1990, *Global Forest Resources,* London: Belhaven Press, 341pp.

Malingreau J.P and Tucker, C.J., 1988, Large scale deforestation in the Southeastern Amazon Basin of Brazil, *Ambio,* **17**, 49-55.

Malingreau, J.P., Tucker, C.J. and Laporte, N., 1989, AVHRR for monitoring global tropical forestation, *International Journal of Remote Sensing,* **10**, 855-68.

Myers, N., 1989, *Deforestation Rates in Tropical Forests and their Climatic Implications,* London: Friends of the Earth, 116 pp.

Nelson, R. and Holben, B.N., 1986, Identifying deforestation in Brazil using multi-resolution satellite data, *International Journal of Remote Sensing,* **7**, 429-48.

Pielke, R.A., Dalu, G., Snook, J.S., Lee, T.J. and Kittel, T.G.F., 1991, Nonlinear influence of mesoscale landuse on weather and climate, *Journal of Climate* (submitted).

Salati, E. and Vose, P.B., 1984, Amazon Brazil: a system in equilibrium, *Science,* **225**, (4658), 129-37.

Singh, K.D., 1990, Design of a global tropical forest resources assessment, *Photogrammetric Engineering and Remote Sensing,* **56**, 1343-52.

Teillet, M. (Ed.), 1990, *Report on a Special Meeting on AVHRR Data Processing and Compositing Methods.* Ottawa: Canada Centre for Remote Sensing.

TREES, 1990, *Tropical Ecosystem Environment Observations by Satellites,* Joint Project of the Commission of the European Communities and the European Space Agency, SP-I.90.31, Ispra: Joint Research Centre.

TREES, 1991, *Strategy Proposal 1991-1992,* TREES Technical Series A, No. 1, Ispra: Joint Research Centre.

Tucker, C.J., Townshend, J.R.G. and Goff, T.E., 1985, African landcover classification using satellite data. *Science,* **225**, 369-75.

Verstraete, M.M. and Pinty, B., 1991, The potential contribution of satellite remote sensing to the understanding of arid land processes, *Vegetatio,* **91**, 59-72.

Woodward, F.I., 1987, *Climate and Plant Distribution,* Cambridge: Cambridge University Press.

World Resources Institute, 1990, *World Resources 1990-91,* Oxford: Oxford University Press.

Chapter 11
Achievements and unresolved problems in vegetation monitoring

M.D. STEVEN, T.J. MALTHUS[1] and J.A. CLARK[2]

Department of Geography,
University of Nottingham,
Nottingham, NG7 2RD

Abstract

Exploitation of the distinctive spectral signature of living vegetation in the visible and near-infrared wavebands has allowed great advances to be made in applying remote sensing techniques to agricultural and ecological monitoring. Applications include the routine monitoring of crops to predict their economic yields, regional early warning systems for famine or pest infestation and phenological mapping of natural vegetation. Many of these successes are based on relating spectral indices of vegetation to the fraction of photosynthetically active radiation (or light) absorbed by the canopy, allowing vegetation indices to be used as direct measures of primary productivity.

When these principles are applied to vegetation that is under stress, a number of problems are encountered. In stressed vegetation, the efficiency of utilization of light absorbed by the canopy may be reduced, causing errors in productivity estimates. Our work has identified a number of spectral indices that may be used to identify when stress is present, but identifying the cause of the stress raises other problems, both of finding a suitable index and in the logistics of deploying it. Firstly, the information content of the data obtained by remote sensing may not be precisely what is required; secondly, the required data may not be available on the appropriate spatial scale or timescale. Moreover, responses of vegetation to stresses are as yet rather poorly understood, so that even knowledge of the type and degree of stress does not imply certain knowledge of its effects on productivity.

Current developments in sensor technology have the potential to enable improved accuracy in the remote sensing of vegetation and its productivity.

---

[1]Now at Wolverhampton Polytechnic.
[2]Department of Physiology and Environmental Science.

New detector systems include imaging spectrometers, which allow many of the biophysical characteristics of canopies to be discriminated, imaging radars, which respond to the structure of foliage, and fluorescence techniques, which may ultimately be able to provide a direct measure of the activity of the photosynthetic system. However, future progress in vegetation monitoring will also depend on improvements in the logistics of operational deployment to match the dynamics and spatial variability of growing vegetation.

Introduction

The spectral signature of vegetation

Considerable advances have been made in the application of remote sensing to monitor vegetation for both agricultural and ecological purposes. The basis for almost all the remote sensing techniques currently applied is the unique 'spectral signature' of vegetation in the optical domain. This spectral signature (Figure 11.1), that is, the characteristic reflectance spectrum of foliage, is a result of two main factors: firstly, the cellular microstructure of a leaf is a very efficient medium for scattering light, producing a characteristic plateau of high reflectance (about 50 per cent) in the near-infrared wavelength region from 750–1200 nm; secondly, in the visible region, absorption by chlorophyll and other pigments, enhanced somewhat by this scattering process, produces a much lower reflectance of 5–10 per cent in the visible region (Gausman, 1985; Guyot, 1989). In remote sensing, reflectance measurements which exploit the contrast between these two parts of the spectrum are often combined in a ratio to form a 'vegetation index' which can be related to various measures of the density of foliage (Curran, 1980; Steven, 1985). A third feature of the spectral signature in Figure 11.1, is the gradual decrease in reflectance in the infrared region beyond about 1000 nm, accompanied by a series of progressively deeper troughs in the reflectance curve. These features are due to absorption by the water in leaves.

In applying the principle of the vegetation index, remote sensing has been anticipated by nature. Plants themselves have light sensitive systems based on the pigment phytochrome which is sensitive to the spectral quality of the light environment. Unlike the photosynthetic apparatus based on chlorophyll, which responds to the total quantity of light available, phytochrome responds to the ratio between the intensities of red light at around 660 nm and near-infrared light at around 730 nm (e.g. Smith, 1976). In plants, this contrast is used to detect the presence of competitors, whose shade is relatively rich in near-infrared light, the visible light having already been filtered out by their leaves. The shaded plants are then able to adapt their morphology, shading inducing extension growth to 'fight for light'.

Light absorption and productivity

Many of the more recent applications of the red/infrared reflectance ratio in remote sensing relate a vegetation index to the fraction of sunlight absorbed by a vegetation canopy. This fraction was shown by Monteith (1977) to be the main determinant of vegetative productivity. In Monteith's analysis, the rate of growth produced by photosynthesis dW/dt is expressed as the product of the incident photosynthetically active radiation (S), the fraction (f) absorbed by the canopy and a conversion quotient (e), usually incorrectly defined as conversion efficiency of absorbed radiant energy into biomass, i.e.

$$dW/dt = efS \qquad (1)$$

In general, S is easily measured by conventional means, although it can also be estimated by remote sensing if required. In principle e is more difficult

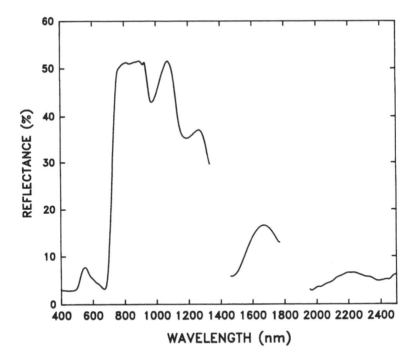

Figure 11.1 Spectral reflectance curve of a sugar beet canopy, relative to a barium sulphate panel. The gaps in the data are caused by water absorption bands and represent areas unavailable to remote sensing.

to assess, but has been found to be relatively invariant when averaged over a growing season, except when the vegetation is under stress (Steven *et al.*, 1990). Relationships between vegetation indices and f are based on the fact that both are largely determined in similar ways by the area and geometry of leaves. These ideas were given a more rigorous theoretical basis by Sellers (1985) and Kumar (1988) and have been tested in the field by many workers (for example, Steven *et al.*, 1983; Asrar *et al.*, 1984; Goward *et al.*, 1985; Prince and Tucker, 1986; Prince, 1990). Therefore, given a healthy green vegetation canopy and an estimate of f by remote sensing, Equation (1) can be used to estimate rates of production.

Applications

Crop yield prediction

One of the main areas of application of this principle is in the prediction of the economic yield of agricultural crops. This was one of the early objectives of the LANDSAT programme and has received a renewed impetus from the requirements of governments to monitor crop production for purposes such as taxation in China (Liu and Zheng, 1990; Li, 1990) or controlling the European commission agricultural budget (King and Meyer-Roux, 1990). The same methodology has also been applied to national monitoring of an individual crop, sugar beet, by Jaggard and Clark (1990). Yield prediction in this system requires the integration of Equation (1) throughout the growing season to estimate the total biomass production, which can be achieved using a sequence of measurements of the vegetation index to estimate the growth curve. Baret *et al.* (1988) suggest that three to five measurements are required, according to the complexity of the model used to fit the curve; Jaggard and Clark use four points. As the yield is generally only part of the total production, the result is multiplied by a known harvest index fraction, according to the crop. In the work reported by Jaggard and Clark, the prediction errors were 0.7 per cent and 5.3 per cent over two years.

Land-cover classification

Equally challenging problems have been encountered in studies of land-use classification. The mapping of vegetation poses considerable problems for remote sensing, even in agricultural areas where distinct crops can be identified, because the spectral signature does not always provide adequate information on its own (Mather, 1990). In part, this is because different types of vegetation have similar broad-band spectral properties at similar stages in the growing season (Paris, 1990). In natural and semi-natural vegetation, the problems are exacerbated because land classes must necessarily be imposed on a continuum of vegetation types. A novel approach to this problem was adopted by Lloyd (1990) who used a year of NOAA AVHRR data to classify vegetation on a phenological basis. A decision tree was used to classify terrestrial vegetation in terms of productivity, based on the timing, duration and intensity of the growing season as defined by the normalized difference vegetation index (NDVI), interpreted as previously outlined.

Early warning systems

The duration of greenness has been recognized as one of the more effective measures of rainfall in semi-arid regions (Rosema, 1990), and such data can have enormous value in sparsely populated areas. For example, desert locusts only breed and swarm when rain provides an adequate supply of foliage, and systems based on measurements with the NOAA AVHRR instrument have been established to monitor desert vegetation and locust breeding conditions. Hielkema *et al.* (1986) established a potential breeding activity factor based on a weighted average of NDVI values in a region. Locust monitoring by such techniques is now performed operationally by the Food and Agriculture Organisation in Rome (Hielkema, 1988) using the ARTEMIS system (Automated Real Time Environmental Monitoring Information System). The same kind of information is also provided to international relief agencies to provide early warning of crop failure.

Problems of stress

Effects of stress on productivity

When remote sensing is applied to vegetation that is under stress, a number of problems are encountered. Stress factors may induce a wide variety of responses in vegetation. The response, if significant, is properly termed the *strain* on the vegetation, which may be broadly classified according to the model of production described earlier. Where stresses cause a response they may result in a reduction in the interception of photosynthetic radiation, or a reduction in the 'efficiency' of conversion, or both. In addition, the harvest index may also be affected. Changes in the first factor pose no new problems for remote sensing but, as illustrated in Figure 11.2, changes in conversion efficiency due to stresses can take a number of forms, each of which introduces difficulties (Steven *et al.*, 1990). Chlorosis is measurable by remote sensing, although the conventional vegetation index confounds this information with the density of foliage. However, stresses which cause closure of the stomata such as lack of water, are less easily monitored, although their effects on canopy temperature are well known (Idso *et al.*, 1982). For example, it was found by Steinmetz (1990) that the computed ratio of actual to potential evaporation determined by remote sensing, accounted for about half the variance in efficiency, enough to indicate that the scientific approach was realistic but giving too weak a relationship to exploit for operational use. In addition, other stresses such as disease may block the photosynthetic apparatus without visible symptoms in leaf pigmentation or effects on stomata. Such problems require a different approach.

A stratified approach

Operational remote sensing requires a stratified approach to the problems of stress. Remote sensing techniques can offer information at the following levels (in increasing order of importance and difficulty):

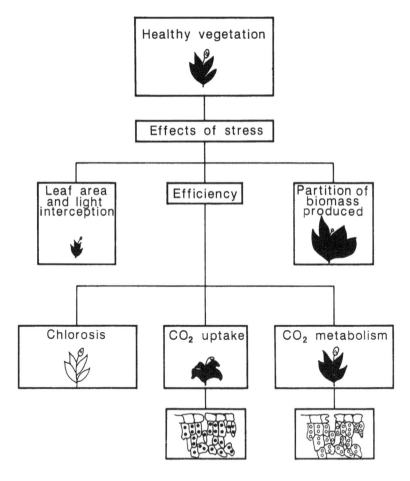

Figure 11.2 Schematic model of vegetation responses to stress. Counter-clockwise from top: healthy vegetation; reduction in leaf area and light interception; loss of chlorophyll (represented by white leaves); reduction of CO$_2$ uptake by closure of stomata; restrictions to the uptake and metabolism of CO$_2$ within the cells; changes in the partition of biomass produced, leading to a reduction in the harvestable component. Reproduced with permission from Butterworths, UK.

(1) the presence of stress (if resulting in strain),
(2) the cause of stress,
(3) the degree of stress,
(4) the effect of stress.

The presence of stress is in most cases already well covered by existing techniques. Most stresses manifest themselves by causing a reduction in leaf area which can be detected by conventional vegetation indices. Another common response to stress is chlorosis (with or without accompanying changes in efficiency), which often occurs before a reduction in leaf area and is easily detected in the visible reflectance. However, establishing the cause of the stress by remote sensing is more difficult. Multispectral aerial photography can be used in some cases, e.g. to detect crop diseases, but the analysis is based more on the spatial pattern than on the spectral reflectance *per se* and satellite data offer neither the spatial resolution required (about 1 m) nor the necessary frequency of sampling to be used in this way (Blakeman, 1990). Measuring the degree of stress is possible in certain cases, e.g. chlorosis, but this information is not always sufficient to determine the effects of the stress, as chlorosis can be caused by a variety of factors whose effects on efficiency do not correspond (Steven *et al.*, 1990).

Indices for stress

As stated above, the conventional near infrared:red vegetation index, when used as an index of foliage density, can give erroneous estimates if stresses such as chlorosis are present in the crop. This is because many stresses affect the plant's ability to absorb and utilize radiation in the red region of the spectrum. Similarly, variations in the colour of the underlying soil adversely affect this index, restricting its application for quantitative purposes to regions of similar soil type.

It seems highly desirable, therefore, to have an alternative index of foliage density that is insensitive to changes in foliage colour and background soil brightness. Research at the University of Nottingham, in collaboration with AFRC Brooms Barn Experimental Station in Suffolk and L'Institut National de la Recherche Agronomique in France, has focused on looking for alternative spectral indices for the independent estimation of foliage cover and plant stress. As an example, we have used high spectral resolution reflectance data to select candidate indices for the estimation of canopy biomass in the presence of canopy stresses and changes in soil background. Figure 11.3 shows changes in the near-infrared to red ratio with canopy cover in an experiment in which reflectances were measured from sugar beet grown at a wide range of densities. In addition to varying the plant densities, half of the plots were deliberately infected with virus yellows disease, while soil background was also manipulated by sliding trays of dark peat between the rows of plants. Although showing the expected strong relationship with percentage canopy cover, the conventional vegetation index was also clearly influenced both by the background soil colour and by chlorosis (caused by the disease). In contrast, alternative indices based on ratios of reflectances in near- and middle-infrared regions of the spectrum showed similar good correlations with foliage cover while being relatively insensitive to the variations in vegetation colour (Figure 11.4). Some of these candidate indices also show less influence of soil background brightness than conventional indices.

The advent of high spectral resolution reflectance measurements, although vastly increasing the volume of data collected, makes feasible alternative methods to detect vegetation to be investigated. One such technique is the use of spectral derivatives as vegetation indices (Demetriades-Shah *et al.*, 1990). The first derivative is the rate of change or slope of the original, zero order, reflectance spectrum. Figure 11.5 shows the relation between the first derivative at 1120 nm and canopy cover, from the same Brooms Barn reflectance experiment as before. This index was as highly correlated with canopy cover, while insensitive to the effects of canopy chlorosis and less sensitive to soil background than the near infrared:red ratio.

Stresses which affect canopy geometry, for example wilting caused by water stress, may also adversely affect the ability of the near infrared:red index to estimate foliage density (Danson *et al.*, 1990). Figure 11.6 shows a comparison of the ratio of reflectances at 800 and 970 nm versus time, between an irrigated and a severely wilted sugar beet crop. The ratio is higher for the irrigated crop, with both plots showing a continuous decrease over the period of measurement, reflecting the changes in canopy water content. It may be difficult to separate the two responses to water stress. Either could provide an adequate signal but the timescale of changes is often in hours, unresolvable on the timescale of all but the weather satellites.

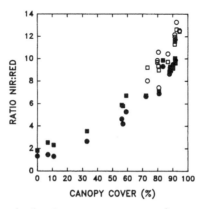

Figure 11.3   Near-infrared:red ratio versus canopy cover for sugar beet (circles, natural soil background; squares, peat background; open symbols, healthy canopy; filled symbols, chlorotic canopy).

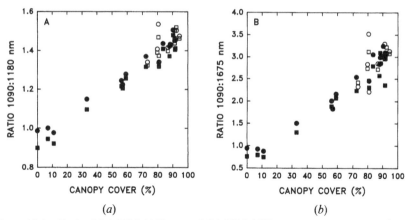

Figure 11.4   Ratio of (a) 1090:1180 nm and (b) 1090:1675 nm versus canopy cover for sugar beet. Symbols as for Figure 11.3.

In summary, vegetation indices based on high spectral resolution reflectances in the near- and middle-infrared region of the spectrum, offer potential for monitoring changes in crop biomass independent of stress-induced effects on reflectance. Similarly, there appear to be suitable candidate indices to separate monitoring of canopy size and stress effects. Further work is required to ensure sensitivity to background soil brightness is kept to a minimum.

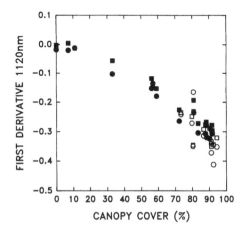

Figure 11.5   First derivative of canopy reflectance at 1120 nm versus canopy cover for sugar beet. Symbols as for Figure 11.3.

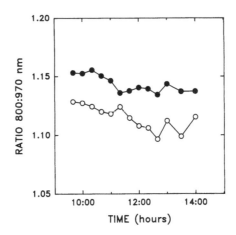

Figure 11.6   Change in ratio of 800:970 nm over time for a wilted (open symbols) and an irrigated (filled symbols) sugar beet plot.

Future techniques

The next generation of instruments to be deployed in Earth orbit include imaging spectrometers such as HIRIS (High Resolution Imaging Spectrometer), with about 250 spectral channels and a spatial resolution of 30 m. As previously discussed, there is considerable potential for monitoring stress effects on vegetation with such instruments, but the high rate of data production imposes restrictions on the area which can be covered or the frequency of repetitive sampling or both (Allan, 1990). Imaging radars are well established and are also due to become fully operational. They offer the advantage of immunity to cloud cover, so allowing reliable, regular monitoring. In principle, radar signals depend on the structure of vegetation canopies and the moisture contents of both the vegetation and the soil. However, the information content of radar images is not easily interpreted in terms of the needs of vegetation monitoring, and further research is required on this application. The solution to reliable monitoring of vegetation may lie in the combination of radar data with data from the more familiar optical domain, as proposed by Paris (1990).

A further potential technique that has been much promoted is the application of fluorescence spectroscopy in remote sensing. Fluorescence by chlorophyll, essentially the emission of unused light energy, occurs whenever the photosynthetic system is restricted, by whatever cause, and unable to use all the light captured. A reduction in the uptake of carbon dioxide is therefore directly related to an increase in fluorescence which can, in principle, give a direct measure of efficiency. Such a technique could therefore detect conditions such as water stress, that are not easily measured by other methods (Methy *et al.*, 1991). Fluorescence techniques offer an early detection of stress effects in the laboratory or on single leaves in the field (Lichtenthaler and Rinderle, 1988). However, a problem that has yet to be addressed is how the principle could be applied to remote sensing. The standard methods for measuring fluorescence in the laboratory involve the kinetics of induction measured over several minutes, usually on dark-adapted leaves, and are not amenable to adaptation for daylight measurements from distant and moving platforms.

Conclusions

Considerable advances have been made in applying remote sensing techniques to the problems of monitoring vegetation, and there is an increasing emphasis in research on developing an understanding of the underlying processes and on using remote sensing to study the dynamics of vegetation on the global scale. Older remote sensing measures of vegetation, such as the NDVI, are being interpreted in a more sophisticated way, and new technologies offer considerable potential for solving problems such as the detection of the effects of stress. However, vegetation systems, being dynamic, are chronically under-sampled by remote sensing techniques, and future progress in vegetation monitoring will depend equally on improvements in the operational deployment of sensors to match the seasonal changes of growing vegetation and its spatial variability.

References

Allan, J.A., 1990, Sensors, platforms and applications; acquiring and managing remotely sensed data, in Steven, M.D. and Clark, J.A. (Eds.), *Applications of Remote Sensing in Agriculture*, London: Butterworths. 3–18.

Asrar, G., Fuchs, M., Kanemasu, E.T. and Hatfield, J.L., 1984, Estimating absorbed photosynthetic radiation and leaf area index from spectral reflectance in wheat, *Agronomy Journal*, **76**, 300-6.

Baret, F., Guyot, G., Teres, J.M. and Rigal, D., 1988, Profil spectral et estimation de la biomasse, *Proceedings of the 4th International Colloquium on 'Spectral Signatures of Objects in Remote Sensing'*, Aussois, France, 18–22 January 1988, European Space Agency, SP-287, Paris, France, 93-8.

Blakeman, R.H., 1990, The identification of crop disease and stress by aerial photography, in Steven, M.D. and Clark, J.A. (Eds), 229–54.

Curran, P.J., 1980, Multispectral remote sensing of vegetation amount, *Progress in Physical Geography*, **4**, 315–41.

Danson, F. M., Steven, M. D., Malthus, T. J. and Jaggard, K. W., 1990, Spectral response of sugar beet to water stress, in Coulson, M. G. (Ed.), *Remote Sensing and Global Change*, Proceedings of the 16th Annual Conference of The Remote Sensing Society, University College, Swansea, 19–21 September, 1990, pp. 49–58.

Demetriades-Shah, T. H., Steven, M. D. and Clark, J.A., 1990, High resolution derivative spectra in remote sensing, *Remote Sensing of Environment*, **33**, 55-64.

Gausman, H.W., 1985, Plant leaf optical properties in visible and near-infrared light, *Graduate Studies*, **29**, Lubbock, Texas: Texas Technical University.

Goward, S.N., Tucker, C.J. and Dye, D.G., 1985, North American vegetation patterns observed with the NOAA-t advanced very high resolution radiometer, *Vegetatio*, **64**, 3-14.

Guyot, G., 1989, Signatures Spectrales des Surfaces Naturelles (Spectral Signatures of Naⅼural Surfaces). *Paradigme*, Caen, France, 178 pp.

Iⅼielkema, J.U., 1988, Use of satellite remote sensing for desert locust survey forecasting at FAO, in *Report on Desert Locust Research—Defining Future Research Priorities*, Rome: United Nations Food and Agriculture Organisation.

Hielkema, J.U., Roffey, J. and Tucker, C.J., 1986, Assessment of ecological conditions associated with the 1980/81 desert locust plague upsurge in W. Africa using environmental satellite data, *International Journal of Remote Sensing*, **7**, 1609-22.

Idso, S.B., Reginato, R.J. and Farah, S.M., 1982, Soil- and atmosphere-induced plant water stress in cotton as inferred from foliage temperatures, *Water Resources Research*, **18**, 1143-8.

Jaggard, K.W. and Clark, C.J., 1990, Remote sensing to predict the yield of sugar beet in England, in Steven, M.D. and Clark, J.A. (Eds), *Applications of Remote Sensing in Agriculture*, London: Butterworths. 201-6.

King, C. and Meyer-Roux, J., 1990, Remote sensing in agriculture: from research to applications, in Steven, M.D. and Clark, J.A. (Eds), *Applications of Remote Sensing in Agriculture*, London: Butterworths. 377-95.

Kumar, M., 1988, Crop canopy spectral reflectance, *International Journal of Remote Sensing*, **9**, 285-94.

Li, Y., 1990, Estimating production of winter wheat by remote sensing and unified ground network. II. Nationwide estimation of wheat yields, in Steven, M.D. and Clark, J.A. (Eds), *Applications of Remote Sensing in Agriculture*, London: Butterworths. 149-57.

Lichtenthaler, H.K. and Rinderle, U., 1988, The role of chlorophyll fluorescence in the detection of stress conditions in plants, *Critical Reviews in Analytical Chemistry*, **19**, 29-85.

Liu, G. and Zheng, D., 1990, Estimating production of winter wheat by remote sensing and unified ground network. I. System verification, in Steven, M.D. and Clark, J.A. (Eds), *Applications of Remote Sensing in Agriculture*, London: Butterworths. 137-47.

Lloyd, D., 1990, A phenological classification of terrestrial vegetation cover using shortwave vegetation index imagery, *International Journal of Remote Sensing*, **11**, 2269-80.

Mather, P.M., 1990, Theoretical problems in image classification, in Steven, M.D. and Clark, J.A. (Eds), *Applications of Remote Sensing in Agriculture*, London: Butterworths. 127-35.

Methy, M., Lacaze, B. and Olioso, A., 1991, Perspectives et limites de la fluorescence pour la teledetection de l'etat hydrique d'un couvert vegetal: cas d'une culture de soja, *International Journal of Remote Sensing*, **12**, 223-30.

Monteith, J.L., 1977, Climate and the efficiency of crop production in Britain, *Philosophical Transactions of the Royal Society of London*, **B281**, 277-94.

Paris, J.F., 1990, On the uses of combined optical and active-microwave image data for agricultural applications, in Steven, M.D. and Clark, J.A. (Eds), *Applications of Remote Sensing in Agriculture*, London: Butterworths. 355-74.

Prince, S.D., 1990, High temporal frequency remote sensing of primary productivity using NOAA AVHRR, in Steven, M.D. and Clark, J.A. (Eds), *Applications of Remote Sensing in Agriculture*, London: Butterworths. 169-83.

Prince, S.D. and Tucker, C.J., 1986, Satellite remote sensing of rangelands in Botswana. II. NOAA AVHRR and herbaceous vegetation, *International Journal of Remote Sensing*, **7**, 1555-70.

Rosema, A., 1990, Comparison of meteosat-based rainfall and evapotranspiration mapping in the Sahel region, *International Journal of Remote Sensing*, **11**, 2299–309.

Sellers, P.J., 1985, Canopy reflectance, photosynthesis and transpiration, *International Journal of Remote Sensing*, **6**, 1335–72.

Smith, H. (Ed.), 1976, *Light and Plant Development*, London: Butterworths.

Steinmetz, S., 1990, *Estimation de l'efficience de conversion en matière seche du rayonnment solaire intercepté par une culture de blé a partir de la réflectance spectrale: relations avec l'evapotranspiration et la température de surface*, Doctorat thesis, Montpellier: Université des Sciences et Techniques du Languedoc.

Steven, M.D., 1985, The physical and physiological interpretation of vegetation spectral signatures, in *Proceedings of the 3rd International Colloquium on Spectral Signatures of Objects in Remote Sensing*, Les Arcs, France, 16–20 December 1985, Paris, France: European Space Agency, SP-247.

Steven, M.D. and Clark, J.A. (Eds), 1990, *Applications of Remote Sensing in Agriculture*, London: Butterworths.

Steven, M.D., Biscoe, P.V. and Jaggard, K.W., 1983, Estimation of sugar beet productivity from reflection in the red and infrared spectral bands, *International Journal of Remote Sensing*, **4**, 325–34.

Steven, M.D., Malthus, T.J., Demetriades-Shah, T.H., Danson, F.M. and Clark, J.A., 1990, High-spectral resolution indices for crop stress, in Steven, M.D. and Clark, J.A. (Eds), *Applications of Remote Sensing in Agriculture*, London: Butterworths. 209–27.

Chapter 12
ERS-1 land and ice applications

C.G. RAPLEY

Mullard Space Science Laboratory,
University College London

Abstract

The European Space Agency's first remote sensing mission, ERS-1, is scheduled for launch in mid–1991. Using an imaging radar, a radar altimeter, and an imaging infrared radiometer (the Along-Track Scanning Radiometer or ATSR), it will provide coverage of the Earth's surface between ±82° latitude with a variety of repeat periods ranging from 3 to 176 days. Its main goals are the observation of the ocean and sea ice. However, ERS-1 will achieve a major advance in the availability of all-weather, day/night radar imagery of land and ice, which will support a wide variety of scientific and commercial applications. Similarly, the radar altimeter and ATSR, although designed for observations of the ocean, will also make valuable contributions to studies of land and ice. The mission is pre-operational, but it and its successor, ERS-2, will lay the foundation for the routine monitoring of the environment into the 21st century by the polar platforms. The establishment of uninterrupted, self-consistent, global data sets is the key to detecting and understanding climatic change.

The ERS-1 mission

The ERS-1 mission represents a major element of the European Space Agency's Earth Observation Programme. Its primary objective is to satisfy both scientific and commercial interests through observations of the Earth's oceans and polar regions. In addition, a high resolution Synthetic Aperture Radar (SAR) is included to achieve all-weather imaging over land and coastal zones.

Specific objectives are to:

(1)   develop and promote applications related to a better knowledge of ocean parameters, sea state and ice conditions; and
(2)   increase the scientific understanding of coastal zones and ocean processes.

However, these far from represent the full mission capabilities.

ERS-1 is characterized by a number of unique and challenging features:

(1) it is defined to be both experimental and pre-operational, since it will demonstrate the technology and design of the space hardware and ground segment, whilst providing an operational capability for certain applications. The number of operational activities is expected to increase during the mission as the experimental work develops;

(2) it is conceived as an end-to-end remote sensing system. In other words, all activities to be performed both on board and on the ground, up to the delivery of data products to users, are an integral part of the system, irrespective of which entities (national or ESA) will be in charge of the various parts of the system;

(3) it is intended to be a *global* mission and to provide world-wide geographical coverage within the constraints of the instrument duty cycles;

(4) specific Fast Delivery (FD) data products will be transmitted to users within three hours or less of data acquisition. On a longer timescale (weeks to months), Off-Line (OL) products will also be available, covering a much wider variety than the FD, and including corrections, calibration data and other information;

(5) instrument calibration both pre- and post-launch will be the responsibility of ESA. They will also assist in the coordination of the scientific community who will carry out extensive surface and airborne campaigns to validate the data products.

Launch will be into a sun–synchronous, 780 km altitude orbit covering the latitude range $\pm 82°$ and is scheduled for mid-1991 (in the event, launch took place successfully on 17 July 1991). The nominal duration of the mission is three years, during which time several different orbit patterns will be executed (Figure 12.1) in order to emphasize specific types of application.

Within this paper a brief outline is given of the instruments ERS-1 will carry, and the associated ground segment, and the wide range of land and ice applications which these will support are summarized. More detailed accounts of the ERS-1 mission and its space and ground segments can be found in the ESA Bulletin ERS-1 Special Issue (February 1991), and in the UK Earth Observation Data Centre's ERS-1 Reference Manual.

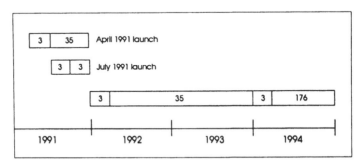

Figure 12.1   The sequence of orbit repeat patterns to be executed by ERS-1. The initial 3-day repeat is to permit rapid returns to calibration sites during the commissioning phase of the mission. Subsequent 3-day repeat cycles are to provide coverage of sea ice break-up in Canadian coastal waters and the Gulf of Bothnia. The 35-day repeat, operated throughout the major part of the mission, allows global SAR sampling and supports ocean circulation studies and sea surface temperature monitoring with the altimeter and ATSR respectively. The 176-day repeat provides close track spacing for the generation of highly detailed altimetric maps of the ocean geoid, ice sheets and land.

Instruments

The ERS-1 satellite weighs 2.3 tons and consists of a spacecraft platform and a payload module (Figure 12.2). The former is derived from the French SPOT programme and provides electrical power, attitude and orbit control, and all command and telemetry support. The latter carries three main instruments, weighs 1000 kg and consumes about 1 kW of power when in full operation. The overall size of ERS-1 is approximately 6 m × 2 m × 2 m, excluding the antennas and solar array, making it the largest ESA satellite yet launched. Once deployed, the SAR antenna is 10 m in length, whilst the solar array measures 12 m × 2.4 m.

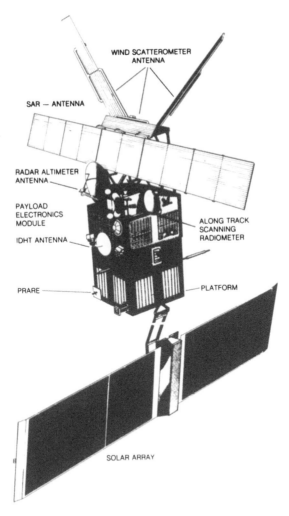

Figure 12.2   ERS-1 satellite showing the instruments and major technical features.

The active microwave instrumentation

The largest of the sensors on board is the Active Microwave Instrument or AMI. This incorporates two separate radar systems, namely the SAR for image and wave mode operation, and a scatterometer for wind mode operation. It is not possible to operate the SAR and scatterometer simultaneously, although operation of the wind and wave modes can be interleaved. Both modes function at C-band (5.3 GHz).

In image mode, the SAR generates radar images over a 100 km wide swath offset about 300 km to the right of the sub-satellite track (Figure 12.3). The incidence angle is $\sim 23°$ (although this can occasionally be altered up to 35° in a special 'Roll-Tilt' mode), and the spatial resolution is 30 m × 30 m. The data rate is 105 Mbps, which is too large to be recorded on board, so the AMI can only be operated when in contact with a suitable ground station. The geographic distributions of SAR ground station coverage, already guaranteed and being negotiated, are illustrated in Figures 12.4a and 12.4b respectively. The heavy power consumption of the SAR further restricts operation to a maximum of 10 minutes per orbit, mostly during daylight.

During the periods when ERS-1 is manoeuvred into a 3-day repeat orbit (commissioning phase and ice phase), the SAR swaths will not overlap except at very high latitudes. A typical coverage pattern over Europe is

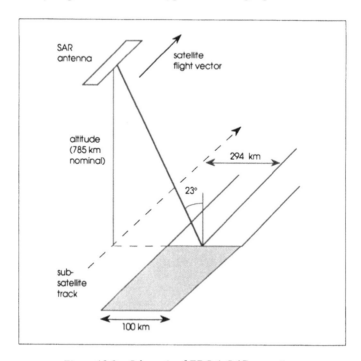

Figure 12.3   Schematic of ERS-1 SAR operation.

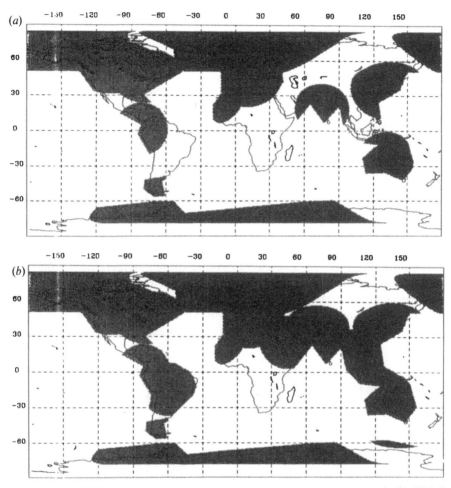

Figure 12.4 (a) ERS-1 SAR coverage with ground stations already negotiated; (b) ERS-1 SAR coverage with all likely ground stations.

shown in Figure 12.5. In the 35-day repeat orbit, full spatial coverage is achievable even at the equator.

The wave mode operation of the SAR provides 5 km × 5 km image 'vignettes' every 200 km, intended for the sampling of ocean wave spectra. The associated data rate is sufficiently low that on-board recording and global operation are possible.

The wind mode uses three antennas to generate beams looking 45° forward, sideways, and 45° backwards to the satellite's flight direction and covering a 500 km swath offset approximately 450 km from the sub-satellite track. The main purpose of this mode is to gather surface wind vector data over the ocean.

Radar altimeter

The radar altimeter is nadir-pointing and is similar in many respects to its highly successful predecessors which flew on the US SEASAT and GEOSAT missions. It operates at Ku-band (13.8 GHz) with a 330 MHz bandwidth in order to achieve a height measurement precision over the ocean of approximately 3 cm rms. Its primary functions are to measure ocean and ice surface height profiles, to measure the power in the echo signal (related to wind speed over the ocean), and to measure the echo waveform shape from which surface roughness can be derived (Figure 12.6). Unlike its predecessors, it has two modes of operation. The ocean mode provides the full height resolution and operates with an on-board tracker system optimized for observations of the ocean. In the ice mode, the bandwidth is reduced by a factor of 4, reducing the height precision correspondingly, but increasing the range of heights recorded in the echo waveform. In addition, a more robust form of tracking is adopted. These two measures, plus the inclusion of an improved acquisition sequence, are intended to ensure more extensive operation over topographic surfaces than has previously been possible.

Associated with the altimeter are a laser retroreflector and the Precise Range and Range Rate Experiment (PRARE). These are used to derive a global orbit ephemeris with a knowledge of the radial component to better

Figure 12.5   ERS-1 SAR coverage over Europe during the 3-day commissioning phase repeat orbit.

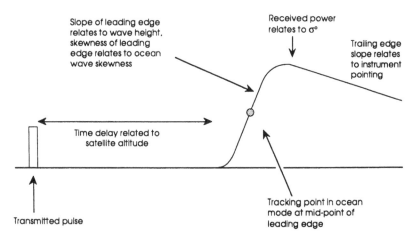

Slope of leading edge relates to wave height, skewness of leading edge relates to ocean wave skewness

Received power relates to σ°

Trailing edge slope relates to instrument pointing

Time delay related to satellite altitude

Transmitted pulse

Tracking point in ocean mode at mid-point of leading edge

Figure 12.6   Schematic of altimeter echo interpretation.

than approximately 50 cm, permitting altimeter profiles to be related to an absolute geodetic coordinate system to a similar degree of accuracy.

The Along-Track Scanning Radiometer and Microwave sounder (ATSR-M)

The ATSR-M consists of an imaging infrared radiometer (the ATSR) and a microwave sounder. The association of the two was for reasons of technical convenience and, in practice, they address different and mainly unrelated objectives. The instrument acquires images across a 500 km swath centred on the sub-satellite track. By using a plane mirror which executes a conical scan, two views of the Earth's surface are obtained, one at nadir and one approximately 900 km ahead, at approximately 55° incidence (Figure 12.7). The pixel size at nadir is 1 km. Images are collected at wavelengths of 11 μm and 12 μm, with a choice of an additional channel at either 1.6 μm or 3.7 μm. Two black body thermal radiation sources are scanned during every mirror rotation to permit very precise and accurate on-board calibration.

The primary objective of the ATSR is to measure global sea surface temperatures to an absolute accuracy better than 0.5 K. This requires an accurate estimation of the atmospheric correction, which is achieved using the multispectral and multi-path-length information.

The microwave sounder points at nadir and operates at 23.8 and 35.6 GHz. Its purpose is to measure the integrated water content of the atmosphere over the ocean, in order to provide refraction corrections for the radar altimeter height measurements.

Ground segment

The ERS-1 mission is the most demanding Earth observation mission yet with regard to data rate, data product generation and delivery. In addition, the scheduling of instrument operation to satisfy both the user requirements and the hardware constraints is particularly complex. An overview of the ERS-1 ground segment is shown in Figure 12.8. The main receiving station is the ESA Kiruna facility in northern Sweden, which acquires 10 of the 14.3 ERS-1 orbits per day. Other ESA stations used include Fucino in Italy, Maspalomas in the Canary Islands, and Gatinau and Prince Albert in Canada. Between them these provide sufficient contact with the satellite to acquire global data coverage from the so-called 'low bit rate' instruments (i.e. wave mode, wind mode, altimeter, ATSR-M, and PRARE), and SAR coverage of Europe and the North Atlantic (Kiruna, Fucino). Additional 'national' (ESA member state) and 'foreign' station reception to extend the area of global SAR coverage has already been referred to (see Figure 12.4). It should be noted that the UK has constructed a SAR receiving station at West Freugh in Scotland, which duplicates part of the Kiruna coverage over Europe and extends the range of imagery over the Atlantic.

A particularly important feature of the ERS-1 mission is the generation and transmission to users of Fast Delivery (FD) data products within 3 hours of observation. The system is designed to provide 3 SAR 100 km × 100 km images per orbit (twice this if no wave mode processing is carried out), wave spectra from up to 200 wave mode ocean scenes, wind mode wind vector

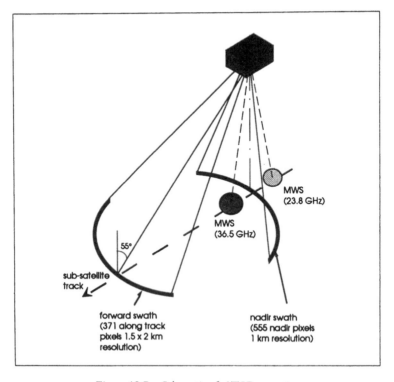

Figure 12.7   Schematic of ATSR operation.

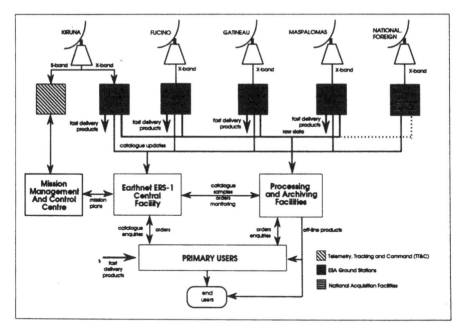

Figure 12.8   Overview of the ERS-1 ground segment.

estimates, and radar altimeter ocean altitudes, wave heights and nadir wind speeds. Currently there are no ESA FD products from the ATSR. The ESA SAR FD data processing is carried out at Kiruna and Fucino, whilst the 'low bit rate' processing is carried out at all the ESA stations.

Although the FD data products satisfy a variety of operational users, the constraints on processing time and availability of supporting data result in their being insufficient for in-depth analysis by non-operational scientific and commercial users. The so called off-line data products are designed to satisfy this wider community.

The generation of the off-line data products is carried out at four Processing and Archiving Facilities (PAFs) provided on ESA's behalf by the UK, France, Germany and Italy. The range of products available is very wide, extending from low-level raw data to high-level geophysical para-meters. The delivery of raw data to the PAFs, the coordination of the PAF tasks and activities, the maintenance of a product catalogue and browse facility, and the handling of user requests (and payments), are all the responsibility of the ESA Earthnet ERS-1 Central Facility (EECF) at Frascati. The EECF also supports a Product Control Service (PCS) responsible for the coordination of algorithm development and software maintenance, quality assessment and control, calibration and validation, and long-term sensor performance assessment. In this way, it is intended to provide effective access to a well-screened and reliable data set, paving the way for future develop-ments such as the Earth Observing System Data and Information System

(EOSDIS) to be implemented later in the decade. However, it should be noted that, with a small number of exceptions, the task of generating thematic data or other types of 'value added' product is the responsibility of the community at large.

Ice and land applications

The AMI imaging mode

The great strength of radar imagery is the ability to operate unaffected by weather (e.g. clouds, fog, rain) or darkness. A disadvantage is the relatively narrow swath, which limits the ability to make repeated observations of a particular region of interest on the Earth's surface. Nevertheless, at mid and high latitudes repeat observations will be possible on shorter timescales than the 35-day orbit cycle during the main part of the mission.

Previous experience with space-borne SARs is limited to the data from SEASAT and the Shuttle Imaging Radars (SIR) A, and B, none of which operated at C-band (all were L-band instruments operating at approximately 1 GHz), and all of which were limited to short lifetimes (three months for SEASAT, a few days each for SIR). None the less, a wide variety of applications over land and ice have been demonstrated and reported both in standard texts (e.g. Sabins, 1987; Elachi, 1987) and the research literature (e.g. Guyenne and Melita, 1985; IEEE, 1986; Wooding *et al.*, 1988).

The situation can be summarized as follows:

(1) *Geology*: The sensitivity of radar images to surface topography, and to surface material dielectric and (especially) roughness properties, makes them ideal for the study of surface structure, geologic age, and geomorphology. Much work has been carried out on the mapping of structural lineaments, especially those associated with faulting and fractures, and on the interpretation of 'radar geologic zones' in terms of surface landforms and rock types (e.g. Sabins, 1987). However, the discrimination of lithological units has proved to be problematical. The ability to penetrate dry superficial deposits to reveal underlying geological and ancient hydrological structures has been demonstrated, although this capability will be reduced at C-band relative to L-band;

(2) *Agriculture*: The main advantage of radar imagery for crop monitoring is the ability to view through clouds, since frequent and reliably timed imaging are crucial for the purposes of crop inventory and yield estimates. The main difficulty concerns the ambiguity between the tones in a single frequency image and a variety of parameters such as crop type, growth state, degree of wind damage, and soil roughness and moisture characteristics. In addition, the tone of certain crops can be sensitive to viewing angle. Future multi-frequency, multi-polarization instruments should resolve these difficulties. However, for the moment additional remote sensing or surface information is generally required. The long-term goal is a reliable method for monitoring land use change, crop type, crop state and crop yield;

(3) *Forestry*: Radar imagery has been shown to be capable of forest cover identification and mapping, and the discrimination of different forest compartments, types, tree species and diseased areas. The primary application is seen as the monitoring of changes in extent, including the identification of clear cut and windblown areas.

Although the all-weather capability is not crucial at mid and high latitudes, where optical imagery can usually be obtained over the normal timescales of change, it is of great value in the tropics, where monitoring of the reduction in extent of the rain forests is a key issue (Figure 12.9).

(4)   *Hydrology*: The difference in backscatter coefficient of water and land, particularly if the water is calm, results in exposed water (lakes, rivers, wetlands) being clearly visible in radar images. The sensitivity of the backscatter coefficient to small changes in surface slope and the presence of windblown deposits, results in dry wadis showing as high contrast features in arid areas. Thus the hydrologic mapping of both wet and dry remote areas is possible. The monitoring of floods is an obvious application, although this may be compromised by limitations on the revisit time if the flooding event is short-lived.

Further potential applications include the study of spatial and temporal variations in surface moisture, and the mapping of surface wetting due to rain, although the ability to achieve quantitative results has not yet been demonstrated;

(5)   *Snow and ice*: The use of SAR imagery for studies of sea ice is well known and will not be discussed here. However, there is considerable potential for monitoring river and lake ice, with particular interest in observing freeze and break up events and in the location and monitoring of ice jams on major river systems.

Radar imagery also clearly distinguishes between snow-free and snow-covered areas (other than open water or marsh), provided the snow is wet. Dry snow is less easy to distinguish. Nevertheless, there is considerable potential for snow mapping and runoff estimation during the melt season. Very little SAR imagery of ice sheets currently exists, but those images which are available suggest that it is possible to distinguish the ablation zone, surface lakes, and other drainage features. ERS-1 data

Figure 12.9   SAR image from SIR-A showing the effects of deforestation in the Amazon jungle.

for the Greenland and Antarctic ice sheets should provide a major advance in this area;

(6) *Topographic mapping*: The radar imaging mechanism results in an ambiguity between height and across-track position which produces image distortions and, in extreme cases, 'layover'. The correction of such artifacts (geocoding) can be achieved using a suitable digital terrain model. Alternatively, if two views of the same scene are available from different vantage points, stereo techniques can be used to carry out the geometric correction and to extract height information. If the two vantage points are very close, the radar phase information can be used to generate an interference pattern between the two images, and this also may be used to geocode the scene and to extract a height map. The ability to produce high resolution digital terrain models globally is recognized as the key to major new advances in a wide variety of geophysical studies, and interferometric radar imaging is regarded as potentially the most effective way to achieve progress.

With regard to feature mapping, man-made structures often produce strong signals through their action as corner reflectors. However, the problems of speckle noise and non-uniform response to different types of feature, limits the scale of effective mapping to greater than 1:100 000. Wise and Trinder (1987) report that the SIR-B imagery contained only 60 per cent of the features depicted on standard 1:100 000 maps.

The AMI wave and wind modes

The AMI wind and wave modes were designed for operation over the open ocean. However, it is feasible to operate them over land, with the advantage that they could provide global access. The wave mode 5 km square 'imagettes' would sample the surface every 200 km, and could be gathered at regular intervals to study seasonal or inter-annual change (note though that it would not be possible to view the same 5 km surface area on each pass). Similarly, the wind mode radar backscatter data could be used to generate coarse resolution (approximately 50 km) global backscatter maps. An analysis of the SEASAT scatterometer data over land revealed systematic, large-scale global patterns of backscatter (Kennett, personal communication).

The radar altimeter

Pulse-limited radar altimeters, such as the instrument on ERS-1, are designed specifically for operation over flat surfaces. However, experience with SEASAT and GEOSAT altimeter data has shown that they can produce very valuable information over sea ice, ice sheets, inland water, and flat areas of land (see Rapley *et al.*, 1987; Guzkowska *et al.*, 1990).

(1) *Ice sheets and ice shelves*: It is not known whether the Greenland and Antarctic Ice Sheets are growing or shrinking in response to climate change. There is concern that a major area of West Antarctica may be unstable and vulnerable to the disintegration of its surrounding ice shelves, with a consequent 5 m change in global sea level. Satellite radar altimeter data offer the means of measuring and monitoring

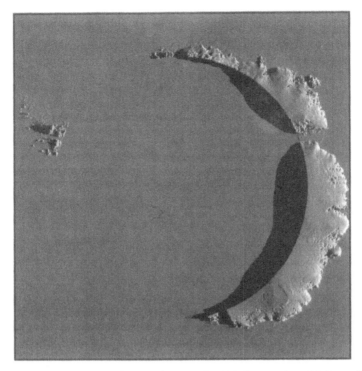

Figure 12.10   Antarctic topographic map from GEOSAT altimeter data. ERS-1 will significantly extend the coverage (courtesy Mantripp).

ice sheet and ice shelf topography to the sub-metre level. This will provide a quantitative basis for investigating ice sheet behaviour. The data also reveal the form and extent of ice shelves, the location of crevassed areas and the climatically-sensitive grounding line, the point at which the ice shelf begins to float. Figure 12.10 shows the result of topographic mapping of Antarctica from the GEOSAT altimeter geodetic mission. ERS-1 will extend the area mapped from 72°S to 82°S, providing coverage of the major Antarctic ice shelves and the climatically-sensitive West Antarctica for the first time;

(2) *Inland water*: The level of water in lakes, wetlands and rivers responds to seasonal and longer-term climatic effects. Of the relatively small fraction of the world's inland water bodies which are gauged, it is possible to obtain data from only a few. The altimeter's ability to measure water levels from a global selection of inland water bodies offers a new means of identifying patterns of change;

(3) *Land surface profiles and topography*: The SEASAT altimeter tracked approximately 34 per cent of the Earth's land surface. The echoes from approximately 10 per cent of the land surface were simple in form and permitted the extraction of surface height estimates to better than 1 m. These profiles reveal subtle slopes and landforms not recorded on standard maps, and provide a basis for evaluating the accuracy of global digital terrain models, which are found to contain substantial errors. A good example is given by the Amazon Basin, where conventional map heights have been found to be systematically in error by up to 30 m (Figure 12.11).

A fundamental limitation of current altimetry is the uncertainty in the radial position of the instrument relative to the Earth's centre of mass (best results to date approximately 30 cm rms). Flat land areas may be used as height reference surfaces to reduce radial orbit error in their vicinity, thereby improving the monitoring of nearby height variable surfaces (e.g. ocean, inland water, ice).

(4) *Transponders:* The use of transponders to obtain accurate height estimates at the ends of short (approximately 2000 km) orbit arcs also permits significantly improved altimetry over the areas in between. The ERS-1 altimeter is the first to incorporate a tracking mode designed to permit 'capture' of transponder echoes. ESA has sponsored the construction of a transponder which will be used to evaluate the technique.

The ATSR

The ATSR represents an advance over the existing NOAA AVHRR infrared imaging radiometers since it can provide global, rather than regional, coverage with 1 km spatial resolution. It also provides a 1.6 μm channel option, and is expected to have lower noise and improved temperature resolution in addition to its greater accuracy. Its one disadvantage is its somewhat narrower swath (500 km). Nevertheless, a variety of land and ice studies are planned. Examples include :

(1) *Lake surface temperatures:* Lakes act as 'integrators' of cloud and air temperature conditions and provide an excellent 'proxy indicator' of local climate. Similarly, the dates of ice freeze-up and thaw for high latitude lakes can also be used to monitor

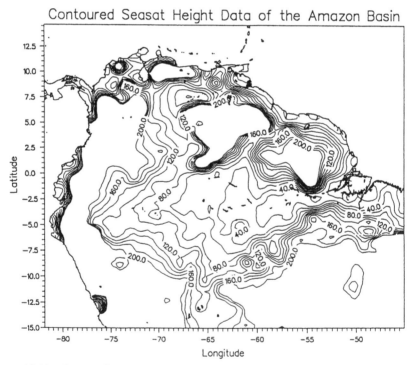

Figure 12.11   Contoured SEASAT altimeter height data of the Amazon basin (courtesy Cudlip).

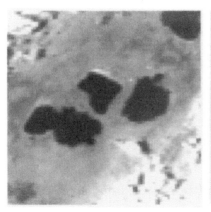

Figure 12.12   AVHRR images two years apart showing a significant change in the surface area of Lake Abiyata (middle of the group of three) in Ethiopia (courtesy Harris).

climatic change. Consequently, a global programme of lake surface temperature studies is planned;

(2)   *Lake and wetland surface areas*: The use of altimeter height data to monitor global lake and wetland level changes is described above. By combining those data with area estimates from the, ATSR the more valuable water *volume* change can be estimated. An example of lake area-change detection using AVHRR data is shown in Figure 12.12;

(3)   *Vegetation 'NDVI'*: The generation of a normalized difference vegetation index (NDVI) has become standard practice with the AVHRR data. The interest in the NDVI parameter derives from seasonal changes in surface emissivity and temperature which reveal global patterns of vegetation-related changes. However, the quantitative interpretation of NDVI in terms of vegetation parameters, such as biomass or plant vigour, has proved elusive. Nevertheless, the ATSR provides the same channels as AVHRR, and so the derivation of an ATSR NDVI will be possible;

(4)   *Ice surface temperatures*: The ATSR data over glaciers and ice sheets will permit the study of seasonal and longer-term variations in temperature and emissivity, of interest in the study of surface processes (freeze, thaw, wind ablation, etc.) and climate change;

(5)   *Altimeter synergism*: One of the difficulties with the interpretation of satellite radar altimeter data in the past has been the lack of supporting data, particularly in remote areas and concerning time-variable phenomena. On ERS-1 the ATSR images the altimeter footprint, providing a means of resolving uncertainties and ambiguities.

Comments and conclusions

There can be no doubt that ERS-1, if technologically successful, will provide a quantum leap in the availability of high quality remote sensing information from the Earth's surface. Over its lifetime, it will collect some 10^{15} bits of radar imagery, 2.5×10^9 surface height estimates, and 2.5×10^{13} bits of infrared imagery.

Will this result in new understanding? Given past experience with remote sensing missions we can certainly say yes; however, the degree to which this will be true will depend on many factors including the effectiveness of the ground segment, the levels of funding for data analysis and interpretation, and the skills of the scientists and other workers involved. The importance of achieving the very highest levels of success cannot be over-emphasized; mankind's ability to achieve a harmonious balance with our environment is the main issue at stake.

References

Elachi, C., 1987, *Introduction to the Physics and Techniques of Remote Sensing*, New York: Wiley.

ESA Bulletin, 1991, ERS-1 Special Issue, Noordwijk, The Netherlands: ESA Publications Division.

Guyenne, D. and Melita, O., 1985, *Thematic Applications of SAR Data*, ESA Special Publication SP-257, Noordwijk, The Netherlands: ESA Publications Division.

Guzowska, M.A.J., Rapley, C.G., Ridley, J.K., Cudlipp, W., Birkett, C.M. and Scott, R.F., 1990, *Developments in Inland Water and Land Altimetry*, ESA-CR 7839/88/F/ FL, Noordwijk, The Netherlands: ESA Publications Division.

IEEE Transactions on Geoscience and Remote Sensing, 1986, Special Issue on Shuttle Imaging Radar (SIR-B). **GE-24**, 4.

Rapley, C.G., Guzowska, M.A.J., Cudlipp, W. and Mason, I.M., 1987, *Exploratory Study of Inland Water and Land Altimetry using SEASAT Data*, ESA-CR 6483/85/NL/BI, Noordwijk, The Netherlands: ESA Publications Division.

Sabins, F.F., 1987, *Remote Sensing: Principles and Interpretation*, San Francisco: Freeman.

UK EODC, 1991, *ERS-1 Reference Manual*, Farnborough: Earth Observation Data Centre, DC-MA-EOS-ED-0001.

Wise, P.J. and Trinder, J.C., 1987, Assessment of SIR-B for topographic mapping, *Photogrammetric Engineering and Remote Sensing*, **53**(11), 1539-44

Wooding, M.G., Sieber, A.J., Churchill, P.N., Lichtenegger, R.J., and Fea, M., 1988, *Imaging Radar Applications in Europe, Illustrated Experimental Results 1978-1987*, ESA Special Publication TM-01, Noordwijk, The Netherlands: ESA Publications Division.

Chapter 13

The EOS Data and Information System (EOSDIS)

J. DOZIER[1]

Universities Space Research Association,
NASA Goddard Space Flight Center,
Greenbelt, MD 20771

Abstract

The Earth Observing System (EOS), the major NASA contribution to the US Global Change Research Program, consists of a series of satellites with sensors designed to measure the crucial variables to monitor processes on the land, ocean, and atmosphere, a Data and Information System (EOSDIS) to analyse geophysical and biological products and distribute them to investigators, and a scientific research programme. EOSDIS begins this year with organization of existing data sets, with the goal of a useful, accessible system in 1994. The launch of the first EOS platform will occur in 1998, with launches every two years to provide a 15-year series of reliable scientific products.

EOS: The Earth Observing System

The Earth Observing System (EOS) is the centrepiece of NASA's Mission to Planet Earth initiative. It is a pivotal part of the US Global Change Research Program (Committee on Earth Sciences, 1990), and hence of the international effort to understand global change and the increasing demands of human activity (Dozier, 1990; NASA, 1990). EOS consists of a space-based observing system, a Data and Information System (EOSDIS), and a scientific research programme. The space component consists of a series of polar-orbiting spacecraft, the first scheduled for launch in 1998, that will collect data for 15 years. EOS is completing its conceptual design phase and is preparing to enter the design phase with the selection and construction of the instruments for the first platform.

---

[1]On leave from Center for Remote Sensing and Environmental Optics, and Department of Geography, University of California, Santa Barbara, CA 93106.

EOSDIS will allow researchers to quickly and easily access data about the Earth system. Development of EOSDIS has already begun; it will support research and analysis with existing data. Geophysical and biological products will be created from the satellite data to be useful to a broad range of the scientific community.

EOS was conceived in the context of a larger venture involving NASA, other agencies of the US Government, and international participants who are providing scientific instruments and data. The space component will be supplemented by European and Japanese platforms and continuing operational and commercial satellites. Additionally, NASA has committed to provide smaller missions, called Earth probes, dedicated to near-term observations of specific Earth processes.

The scientific research programme was initiated in 1990, with funding for 29 interdisciplinary teams, to begin development of models that will use EOS data and define data requirements from the instruments; nine facility instruments and their science teams; 23 instrument investigations; and definition studies for a synthetic aperture radar.

Importance of EOSDIS

Crucial to the success of the Earth Observing System is the EOS Data and Information System (Science Advisory Panel, 1989). The goals of EOS depend not only on its instruments and scientific investigations, but also on how well EOSDIS helps scientists integrate reliable, large-scale data sets of geophysical and biological measurements made from EOS data, and on how successfully EOS scientists interact with other investigations in Earth system science. Current progress in the use of remote sensing for science is hampered by requirements that the scientist understand in detail the instrument, the electromagnetic properties of the surface, and a suite of arcane tape formats, and by the immaturity of some of the techniques for estimating geophysical and biological variables from remote sensing data. These shortcomings must be transcended if remote sensing data are to be used by a much wider population of scientists who study environmental change at regional and global scales.

Evolutionary development of EOSDIS will support research and analysis of existing data, with extensive involvement by scientific users in all phases including studies, requirements definition, design, and testing. EOSDIS development will start with 'Version 0', giving access to existing Earth science and applications data systems at distributed locations, and will support work with currently available data of high interest to scientific users. EOSDIS will then gradually grow in its capabilities to acquire, process, archive, and distribute large volumes of data from EOS instruments as they are deployed.

Scientific information from EOSDIS

What distinguishes EOSDIS from most current remote sensing data systems, is the commitment to provide usable scientific information to the geophysical, biogeochemical, ecological, and interdisciplinary communities. EOS data products will be used by a wide spectrum of scientists and the public during the 15-year life of the Mission, and for decades afterward. Standard, reliable data products, essential to distinguish natural and anthropogenic variations, will give the community access to independent measurements to validate and drive models of processes at local, regional and global scales. The characterization will include algorithms for generating the products and descriptors of data quality, and each data set will include the identity of the responsible scientists. Standard products, created by algorithms that are certified by a peer-review process, will be available through EOSDIS for all cases in which appropriate input data exist. Moreover, specialized products, created by EOS scientists in their own computing facilities, are to be archived and distributed by EOSDIS. Geophysical and biological products will be available in EOSDIS for use by other scientists. In this way, EOS will open the capabilities of remote sensing data to a broader range of the scientific community, who will no longer need detailed knowledge of instrument characteristics and electromagnetic interactions at the surface.

Evolving design and architecture

The important questions that drive the design and architecture of EOSDIS involve the procedures by which scientific products will be created and distributed, as well as how the styles of interaction with EOSDIS will change as EOS science matures. An EOS data processing and distribution system, including visualization and browse capability for both image and non-image data and information derived from EOS sensors, requires an evolutionary, distributed design, due to flexible research utilization of the data and because of rapid developments occurring in computer hardware and software. As EOS matures, specific algorithms will be formulated, mutual product dependencies resolved, and interdisciplinary data requirements defined. The process of producing and analysing data will continue to lead to new methods of producing scientific products and new computing requirements. The system architecture must accommodate data from different kinds of sensors, changes in available computer hardware, software, and communications, different levels of human involvement in the creation of standard products, and different centres of expertise. A system that can address the changing nature of both its tasks and the available hardware and software, inherently must be designed for easy, graceful evolution, both before and after launch of the EOS platforms.

To incorporate evolution into the design of EOSDIS is a challenge. In the design and development of EOSDIS, we must:

(1) develop scenarios for processing of EOS data into scientific products that will address the style, volume, and human interaction needed by the processing facilities;
(2) promote diverse user facilities that take advantage of existing and anticipated scientific expertise;
(3) develop realistic expectations for rapid browsing and visualization of large data sets;
(4) adhere to a flexible, distributed, portable, evolutionary design;
(5) distribute data products by appropriate high-bandwidth communication or other media; and
(6) operate prototypes in a changing experimental environment.

Relationship to Global Change Program

EOS is a pivotal part of the US Global Change Research Program, and hence of the international effort to understand how the Earth functions as a complete system. Earth system science objectives require a data and information system that will ease and encourage multidisciplinary and interdisciplinary investigations. Data from EOS platforms will be combined with data from other agencies and nations, including remote sensing data from other satellites or aircraft and *in situ* operational and experimental data. These data sets will be accessible and integrated with scientists' computing facilities and models of environmental processes and global change.

Although a part of the Global Change Program, EOS has other objectives as well; it cannot and should not take responsibility for all that is necessary to meet the needs of that programme. Nevertheless, the magnitude and scope of EOS are such that it contains within itself a microcosm of most issues, as well as having a myriad of interfaces to the broader programme both within and outside NASA. It is clearly the flagship endeavour, and its strategy for approaching the human issues of information management will guide progress in the programme as a whole.

We must therefore:

(1) provide an environment of incentives and challenges that persuade scientists to devote their efforts to those tasks in EOSDIS for which their creative talents are essential;
(2) establish dependable linkages, that really function as dependable partnerships, with other agencies and nations that collect, archive, and analyse Earth science data; and
(3) establish data and information interfaces between EOSDIS and other national and international active data archive centres for multidisciplinary and interdisciplinary scientific data analysis.

Continuity with present systems

EOS data will continue existing measurements, some of which currently extend for more than a decade. EOS science begins now, not with the launch of the first EOS platform in 1998. Starting immediately, EOSDIS will develop current and previous data sets and measurements, and provide them to other investigators, so that the EOS community can gain experience with data processing, archive, and distribution centres. A few current centres where remote sensing data are intensively and routinely analysed into scientific products, provide the heritage for design and prototyping of EOS data processing and distribution, especially for data sets consisting of scientific interpretations rather than satellite-level radiance measurements.

We will therefore:

(1) identify appropriate current and previous data and promote their rapid development in order that they will be compatible with anticipated EOS data;
(2) assist the EOSDIS designers to acquire experience with data processing centres and archives that are currently active; and
(3) begin the development of data and information interfaces between EOSDIS and other national and international active data archive centres for multidisciplinary and interdisciplinary scientific data analysis.

Functional objectives

The key functional objectives of EOSDIS are:

(1) Command and control of NASA polar platforms—the first platform, carrying about five instruments, is planned for launch in 1998, with launches planned every 18 to 24 months thereafter. Each will have an expected life of five years, so to ensure a 15-year data set, each will be replaced twice;
(2) Command and control of EOS instruments—brief descriptions are given in the EOS Reference Handbook (NASA, 1991). Because of its unique requirements, the EOS SAR will be flown on a separate platform, slated for an independent new funding start in 1993 or 1994 and a 1999 or 2000 launch;
(3) Processing and reprocessing of EOS data—EOSDIS must support the generation of both standard and special data products. Standard products are of wide research utility, are routinely produced by a peer-reviewed algorithm, and are available anywhere the input data are available. Special products are produced on limited subsets of data, by algorithms that may still be under development. At present, the EOS investigators have defined several hundred candidate standard products;
(4) Data archiving, storage, and distribution—EOSDIS must be able to store all computed standard and special products, during the mission life, and distribute requested subsets of them to users. Data from non-EOS sources that are needed for the generation of products will also be available through EOSDIS;
(5) Information management—EOSDIS must provide information about data (metadata) at adequate granularity and richness to permit easy location and selection of data of interest to users, so that they may decide which data to analyse more intensively. Convenient means include user-friendly interfaces and browsing and visualization tools;
(6) Networks—EOSDIS must provide electronic access to data and information, so that scientists can communicate with each other and with the system;
(7) Transfer to permanent archives—at the end of the mission, the data held by EOSDIS should be transferred into the control of permanent archival agencies, namely NOAA and USGS, through sharing of budgets rather than physical movement of data;

(8) Exchange of data, commands, algorithms, etc.—EOSDIS needs to develop inter-
     faces with NOAA, ESA, NASDA, CCRS, and other agencies to exchange data,
     commands, algorithms, metadata, etc.

Policy on availability of data

NASA policy specifies that all EOS data and derived products be available
to all users, with no preference given to EOS investigators and no
proprietary period. Research users in the US and participating countries,
will pay only the marginal costs for data reproduction and distribution;
they will have to agree to publish their results and to make available
supporting information, including methods of analysis and code implementing the algorithms. Research users in other countries may have the
same access to EOS data by proposing cooperative projects and associated
contributions—similar access to their satellite, aircraft, and surface data. For
all data products, the documented scientific software that produced them
will also be available.

To the extent possible, we want to apply the same policy to non-EOS
data. Other US agencies involved in EOS—NOAA and the US Geological
Survey—have agreed. For data from the international platforms, discussions
are under way between NASA headquarters and the appropriate foreign
government agencies. Expectations are that they will agree to the same data
policy. Availability of commercial data (LANDSAT and SPOT) under the
same policy will require a change in legislation. The LANDSAT system, in
particular, has priced data for full cost recovery, but the usage declined
dramatically when this policy was implemented (Don Lauer, personal
communication).

System architecture

Reasons for distribution of EOSDIS functions

Separation of product generation into different nodes within EOSDIS is
consistent with scientific goals and interdisciplinary research. In order for a
networked information management system to succeed, standards for
operating systems and data formats are crucial, and a single common
catalogue is needed. Any user should be able to investigate the availability
and characteristics of all archived data, without having to use separate
catalogues for different instruments or to learn new access techniques.

Scientific functions of EOSDIS—distribution of instrument Level 1 data
to investigators, information management, interaction among investigators,
creation of geophysical and biological products, and archiving and distribu-

tion of data and information—should be separately optimized. Smooth interfaces are also important.

A major consideration in arriving at the distributed architecture is the existing reality of a distributed community of investigators and resources. EOSDIS will develop in an environment that is distributed. Proposed EOSDIS related research facilities with associated computational resources are distributed across the continental US and beyond. The investigators are broadly geographically distributed. Further, the global change community that will be a major user of EOSDIS is distributed.

This geographic distribution of facilities and investigators does not allow advantage to be taken of the great strengths of these investigators' skill and excellence. Indeed their distribution, without adequate attention to the requirements of scientific interaction, has not served to extract the potential of research enterprise. EOSDIS must become the logical integrator of facilities by modifying infrastructure and removing roadblocks to effective remote use. It must be possible to remotely access and acquire data that are stored in widely disparate forms, thereby reducing the effort needed before such data can be used in computer analyses. Networks for use in a distributed environment, such as EOSDIS, must evolve in capability, capacity and ubiquity. Individualized styles of research must be accommodated.

The greatest potential for knowledge from EOS comes from the fertilization that follows from interactions among researchers with differing views of data and different styles of data use. This fertilization requires that communication mechanisms improve and that EOS become a driving force in the initiation of the 'laboratory without walls.' It must be far simpler to use and acquire data than it is now. EOSDIS must provide access to data and tools that allow the solution of problems in understanding the Earth as a system. Existing centres of expertise and excellence should be incorporated into a distributed EOSDIS architecture, to allow the best availability and interaction among all resources. A centralized view of EOSDIS assumes that one group has a monopoly on knowledge of how to manage Earth observation data.

Communication among nodes

A challenge of EOSDIS will be to augment communications infrastructure, to further interconnect the research community to itself and the various EOS elements. This augmentation has to occur whether EOSDIS is functionally distributed or centralized. It is more likely to be effective if it is realized at the onset that this augmentation is a key to EOSDIS success, instead of something that is merely an expensive frill to be added when funding is not tight.

EOSDIS is not an isolated system; it exists to help the research community flourish and return useful knowledge about the Earth. It cannot provide resources only to the small segment of the community implied by centralization, but instead must assure that it encourages participation by as broadly based a community as possible.

During the first three decades of satellite observations of the Earth, we learned to put reliable instruments in space and began to make some progress towards understanding how to use these new tools of understanding. However, we regarded data as a precious resource and hoarded them jealously. Each research group that obtained satellite data put them into a form that suited their own experience and current needs. As a result, these data are fragmented and dispersed.

Specific functions

The architecture of EOSDIS has evolved over the past three years though design studies and interaction with the science community. There are three segments to EOSDIS:

(1) The Flight Operations Segment controls the platform and instruments, supports mission planning and scheduling, and monitors health and safety of instruments. It consists of the EOS Operations Centre, Instrument Control Centres, and Instrument Support Terminals.

The EOS Operations Centre coordinates EOS platform and instrument operations and monitors the accomplishment of mission objectives. It also maintains health and safety of the observatories, supports planning and scheduling of the resources on the EOS platforms, coordinates observations from all instruments to develop conflict-free schedules, accommodates unplanned schedule changes, and develops and implements contingency plans. It will normally have to support simultaneous operations of two US platforms, but during replacement of platforms it will have to support simultaneous operation of three platforms over a nominal overlap period of about six months. It receives commands from the Instrument Control Centres, performs high-level command validation to ensure that there are no conflicts, and merges instrument and platform commands for total observatory operations. The EOS Operations Centre also coordinates with the mission operations centres for the European and Japanese platforms and for Space Station Freedom.

EOSDIS has two Instrument Control Centres, one at NASA GSFC and one at JPL. Their functions are to plan and schedule instrument operations, generate and validate instrument command sequences, forward commands in real time or store for later transmission, and monitor health and safety of instruments. They create instrument-specific commands within the schedules provided for each instrument by the EOS Operations Centre, and they review quick-look engineering and science data.

Instrument Support Terminals are provided to the instrument principal investigators and facility instrument team leaders to help monitor instrument status;

(2) The Science Data Processing Segment is the part of EOSDIS of most interest to the investigators. The current concept of EOSDIS is that several 'Distributed Active Archive Centres' (DAACs) will fulfil all processing needs except algorithm development and individual scientists' investigations. At present, seven DAACs have been designated by NASA, and there are Affiliated Data Centres (ADCs)

with which EOSDIS has interfaces. The seven DAACs and the presently identified ADCs are:

Distributed Active Archive Centres: Alaska SAR Facility, Fairbanks; Jet Propulsion Laboratory, Pasadena; NASA Goddard Space Flight Center, Greenbelt; NASA Langley Research Center, Hampton; NASA Marshall Space Flight Center, Huntsville; National Snow and Ice Data Center, Boulder; US Geological Survey, EROS Data Center, Sioux Falls.

Affiliated Data Centres: Consortium for International Earth and Science Information Network, Ann Arbor, University of Wisconsin, Madison; NOAA.

Each will have a 'Product Generation System' (PGS) that will generate standard products, a 'Data Archive and Distribution Systems' (DADS) that will distribute data sets to investigators, and that will be accessed by an Information Management System (IMS). Explicit in this concept is that multiple facilities and generic classes of facilities can best fulfil these functions. The ADCs will not have responsibility for generation of level-2 geophysical or biological products from EOS data, but instead will organize large data sets from other sensors and incorporate large-scale models. In the next year, each potential DAAC will identify appropriate current and previous data and promote their rapid development. They will also acquire experience with currently active data processing centres and archives and begin development of interfaces between DAACs and between EOSDIS and other national and international archives.

The Product Generation System is responsible for the generation of standard data products. The combined capacity at the PGS nodes must be great enough to generate all standard products at a rate fast enough to cope with the incoming data stream and to allow for reprocessing. Algorithmic software for product generation is designed and implemented by the responsible scientists, who also define contents of metadata and browse products associated with the standard products.

The Data Archival and Distribution System archives instrument and interdisciplinary data products, ancillary data, radiometric and geometric calibrations, metadata, command history, correlative data (including those from pre-EOS sensors, surface measurements, and non-EOS data used in product generation), algorithms and documentation. Total data over a 15-year mission is estimated to be about 11 PB. The data are distributed on request to EOS scientists, other DAACs and the general community, via either electronic networks or on media such as optical discs or magnetic tapes.

The primary function of the Information Management System is to provide information about the data holdings in EOSDIS and access to other (external) archives. The IMS will be distributed, to take advantage of the diversity in experience at the DAACs and to permit DAAC-specific features. The degree of distribution of functionality and the configuration will depend on the state of database management technology and network responsiveness. Regardless of which DAAC a user interacts with, the IMS will provide uniform, seamless access to all data held by EOSDIS, through convenient, easy user interfaces for novices and experts. It will be possible to access data by simple search criteria, such as instrument name, product name, time of collection or spatial coordinates. Moreover, other modes of searching should be provided to permit cross-instrument and cross-disciplinary searches by enriching the metadata with summaries of the data sets. The IMS is the element through which data are ordered by users.

A user support office at each DAAC consists of scientific experts and support staff to assist users in understanding the data products specific to that DAAC. Each works closely with the scientific community through science advisory groups in its own discipline and the EOSDIS Advisory Panel. Activities are coordinated through the EOS Science Processing Support Office;

(3) The Communications and System Management Segment services the DAACs and

the scientists' computing facilities with the connectivity and management functions to ensure appropriate data flows, management of production schedules, and resource usage.

The EOSDIS Science Network, possibly a combination of NASA institutional and other existing or new networks, will electronically distribute data among DAACs and scientists' computing facilities, and with the international community. It is anticipated that at least 45 Mbits/s will be needed among DAACs and from 56 Kbits/s to 1.5 Mbits/s between the DAACs and the scientists.

The Systems Management Centre has as its functions configuration management, high-level scheduling of system, site, and element activities, monitoring production and performance, resolving faults, establishing security, accounting, and billing;

(4)   Field Support Terminals (FST) provide scientists in field campaigns with mobile communications to permit coordination of platform data with field experiments;

(5)   Scientific Computing Facilities (SCF) are the scientists' facilities to develop and maintain algorithms and software for producing scientific products, control quality of the standard products, support data set validation, instrument calibration and analysis, generate special products, and provide needed resources for the scientist's research. A set of software tools is provided to the SCFs to help them interact with each other and with other EOSDIS elements.

Analysis in the EOS era

The Earth Observing System and its Data and Information System offer a chance to view the Earth and data about the Earth from a new, global perspective. If we are wise in what we do, we may be able to create new understanding of how the Earth works as a system through combination of data in new and unexpected ways. Achieving this global view will not be easy. There are many roadblocks; data are stored in widely disparate forms that cannot be used in computer analyses without strenuous effort by researchers, some data are not stored electronically and electronic networks for distributing research findings and data are primitive. Furthermore, communities and individual researchers have highly individual views of data and styles of research. These styles and views have served well during the previous era of research from space, yet, they make simple communication across disciplines difficult, as each group of researchers has its own nomenclature and habits of mind.

We believe that the real potential for knowledge from EOS comes from the fertilization that follows from the interaction between researchers with different views of the data and different styles of interacting with it. This fertilization requires that investigators be able to communicate with each other more easily than they do at present, and that they be able to use data more efficiently. Old habits will not change unless they can be easily replaced with new ones that provide recognizable rewards; researchers need to show the usefulness of new techniques on problems they are familiar with and in settings where they can take maximum advantage of tools and skills they have acquired through hard work and experience.

In order for EOSDIS to be accepted, we must provide researchers with access to data and to existing tools that will allow them to solve current problems in understanding how the Earth works as a system. We cannot expect this to happen by ignoring existing centres of experience and expertise; we will obtain the most efficient use of our resources by applying them to groups that are already proficient in using data.

In short, EOSDIS is not really an independent hardware and software system. It exists to help the research community flourish and return useful knowledge about our threatened planet. We cannot provide a resource to only one part of the research community without harming other parts, particularly by closing avenues of communication that must provide the most useful cross-fertilization.

Conclusion

Starting with the middle of this decade, EOSDIS will be NASA's primary data and information system for Earth science. The success of EOSDIS will be tightly linked to the success of the scientists using the data from EOS and other sources for research in Earth system science, and the significant discoveries made by them. The role of EOSDIS will be to provide easy and quick access to usable, understandable and timely data, and to foster cross-fertilization between disciplines.

Some of the key challenges faced by EOSDIS are: satisfying the data and information needs of a diverse, multidisciplinary scientific community; integrating product generation algorithms for over two dozen instruments; keeping up with an orbital average data rate of over 50 Mbits/s and assuring prompt generation of several hundred standard products; reprocessing data and algorithms change; and storing, distributing, and managing information about tens of petabytes of data over the 15-year life of the mission.

Careful planning and coordination among many organizations and institutions are needed to ensure the success of EOSDIS. The development must be evolutionary because of the long life of the system, the community being served, and the rapid changes in applicable technologies.

EOSDIS will be implemented as a suite of discipline-oriented Distributed Active Archive Centres, and development will start immediately and be evolutionary, based on existing data systems. Existing data sets and data from EOS precursor missions will be used to exercise, prototype, and validate system capabilities and detailed requirements with continued involvement by scientific users of the data.

Acknowledgement

In this summary description of EOSDIS, I have borrowed freely from discussions with my colleagues and from published and unpublished documents circulating among EOS investigators. To name them all and identify their individual contributions is too daunting, and I e apologize here for the many omissions. Specifically though, I must thank my colleagues on the EOSDIS Advisory Panel, at the EOSDIS Project at Goddard Space Flight Center, and at NASA headquarters.

References

Committee on Earth Sciences, 1991, *Our Changing Planet: The FY 1992 US Global Change Research Program*, Washington D.C.: Office of Science and Technology Policy.

Dozier, J., 1990, Looking ahead to EOS: the Earth observing system, *Computers in Physics*, **4**, 248-59.

NASA, 1990, *EOS: A Mission to Planet Earth*, Washington D.C.: National Aeronautics and Space Administration.

NASA, 1991, *EOS Reference Handbook*, Greenbelt, MD: NASA Goddard Space Flight Center.

Science Advisory Panel for EOS Data and Information, 1989, *Initial Scientific Assessment of the EOS Data and Information System (EOSDIS)*, Eos-89-1, Greenbelt, MD: NASA Goddard Space Flight Center.

Chapter 14
The ESA Earth Observation Polar Platform Programme

M. RAST and C.J. READINGS

European Space Agency

Abstract

The Earth's environment and especially the means required to monitor and preserve it, are being increasingly recognized as matters of great concern. Within its Earth Observation Programme, ESA is studying at present the payload and mission configurations of the European Platform which will be dedicated to the study and monitoring of the Earth/atmosphere system.

The provision of remotely-sensed data from polar-orbiting satellites is essential for the acquisition of information needed to address the environmental problems now facing mankind. These data are needed for the following reasons:

(1) to identify and advance understanding of relevant processes;
(2) monitor the state of the Earth/atmosphere system.

The former is tied to the development of models and the latter to the detection of changes (plus the validation of models). These data cannot be obtained without recourse to observations from space; neither can they be provided by one nation working alone. International cooperation is essential.

This paper concentrates on the planned contribution of the European Space Agency (ESA) to Earth observation and environmental monitoring during the Polar Platform era. The overall scenario is outlined, and information is provided on the instruments currently being considered for flight on the European Polar Platforms. This programme is intended to make a significant contribution to the study of environmental problems.

Introduction

More organized information about the behaviour of the environment and the factors influencing the Earth's natural resources can only be achieved on the basis of a better understanding of the Earth when viewed as a system in which the physical, chemical and biological interactions between the

175

atmosphere, the oceans, the land and the ice regions, plus the structure of the Earth itself, are all taken into account.

Fundamental to this is the provision of data to identify processes and validate models. These data are also needed to monitor the state of the Earth system and to detect changes. In most instances long-term continuity of data is also essential (e.g. for climatology, environmental change detection), requiring that spaceborne capabilities be sustained for periods of a decade or more by their replacement at the end of their lives. These needs are addressed in the Earth Observation Programme proposed by the European Space Agency (ESA). It forms the basis of the individual programme proposals which are submitted to the ESA member states for approval.

Underlying the ESA Earth Observation Programme is the Agency's Earth observation strategy. This was enunciated in collaboration with the user community and representatives of the ESA member states. It identifies four clear themes, namely:

(1) monitoring the Earth's environment on various scales, from local or regional to global;
(2) management and monitoring of the Earth's resources, both renewable and non-renewable;
(3) continuation and improvement of the service provided to the world-wide operational meteorological community;
(4) contribution to the understanding of the structure and dynamics of the Earth's crust and interior.

The realization of these objectives forms the basis of the Agency's Earth Observation Programme. It depends on an integrated future programme, one of the main elements of which is a series of Polar Orbit Earth Observation Missions (POEMs). These POEMs are intended to provide the continuity of data in polar orbit that are essential to the realization of many of the basic objectives.

The Polar Orbit Earth Observation Missions

To address the four themes underlying the Agency's Earth observation strategy, global data sets, in many cases spanning decades, are required. These data must cover a wide range of disciplines. Furthermore, given the synergism between both disciplines and instruments necessary to realize objectives, missions will in general have to be broader in scope and carry more instruments than has been the case in the past. A consideration of ways to contribute to the above objectives lead the Agency to conclude that it should implement a series of polar platform missions based on the Columbus polar platform currently under development in ESA.

These platforms have an expected lifetime of about 4.5 years with a payload carrying capability variable between 1000 and about 2400 kg, roughly two to three times that of ERS-1. The first platform should be available for launch in 1997. As an optimum implementation strategy for polar orbit, it is proposed to deploy two series of ongoing missions. This is the so-called dual orbit continuous scenario depicted in Figure 14.1 (the link to ERS-1 and ERS-2 is also illustrated). In this figure two separate lines of polar platforms are distinguished, namely the M series (EPOP-M1, EPOP-M2, etc.) and the N series (EPOP-N1, EPOP-N2, etc). Each focuses on a different set of primary mission objectives:

> M series—primarily meteorology/ocean/climate
> N series—primarily Earth resources/atmosphere

Together they span the objectives of the Earth Observation Programme (Figure 14.1)

The missions will utilize sun-synchronous orbits with a morning descending node and different (optimized) orbit altitudes lying in the range between 700 km and 850 km. They are intended to complement missions planned by other space agencies, notably NASA, which also envisage flying a series of polar platforms.

The payload of the first polar mision

The first ESA polar platform mission, EPOP-M1, is envisaged as a combined meteorology-atmosphere/ocean-ice/climate mission with its payload comprising three main elements, outlined in the following sections.

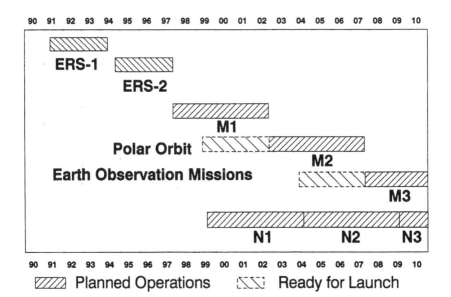

Figure 14.1   Polar Orbit Earth Observation Missions (see text for explanation).

Operational meteorological instruments

In the main, these are improved versions of the operational instruments currently flying on the NOAA Tiros-N polar-orbiting operational meteor-ological satellites:

VIRSR   Visible and InfraRed Scanning Radiometer.
               Radiometer dedicated to global monitoring of clouds, sea surface temperature, vegetation and ice, operating in seven spectral bands over a swathwidth of 2000 km with a 1 km spatial resolution. This instrument is derived from the NOAA AVHRR;
IASI   Infrared Atmospheric Sounding Interferometer.
               Interferometer for improved temperature and humidity sounding and trace gas monitoring. The objective is to achieve an accuracy of 10 per cent in observing integrated column contents of O_3, N_2O, CH_4, CO. This is a new high-resolution sounder;
IRTS   InfraRed Temperature Sounder.
               Scanning infrared radiometer dedicated to measure global atmospheric temperature profiles, atmospheric water content, cloud properties and Earth radiation budget in 20 channels from 0.2 to 15 μm, with a swath width of 2250 km, and 21 km resolution at nadir. (This instrument is derived from the NOAA HIRS.);
AMSU   Advanced Microwave Sounding Unit (Microwave Temperature
(MTS/MHS)   Sounder/Microwave Humidity Sounder).
               Global temperature profile, water vapour, precipitation, sea ice, snow cover, and ocean wind stress. Passive microwave radiometer scanning $\pm 48.5°$ from nadir. Fifteen channels for MTS and five for MHS;
MCP   Meteorological Communications Package.
               Provides direct data handling and broadcast of operational instruments to ground stations;
ARGOS   Data Collection and Location System.
               Relays messages and location data from platforms;
S&R   Search and Rescue Package.
               Monitors distress signals of, *inter alia*, ships and planes on internationally acknowledged emergency frequencies.

It is envisaged that these instruments will be jointly provided by NOAA and EUMETSAT as the means for continuing operational meteorological observations in the morning orbit after 1997, when NOAA only intends to operate spacecraft in the afternoon orbit.

Facility instruments

These are instruments which would be developed by ESA. The current list comprises the following:

SAR   Synthetic Aperture Radar.
               High-resolution (25×25 m) imaging of the land, coastal zones, ocean waves and ice. Operates at C-band (5.3 GHz) under all-weather, day or night conditions;
SCATT   Wind Scatterometer.
               Measures wind speed and direction over the ocean with a 50 km resolution and an accuracy up to 2 m/s. Synergism with RA-2, MERIS

and AATSR for biophysical characterization of the ocean, ocean dynamics and energy exchange;

MERIS   Medium Resolution Imaging Spectrometer.
Pushbroom spectrometer that monitors biophysical parameters of the ocean (e.g. ocean colour) in 15 programmable, 5-10 nm wide spectral bands and over a swathwidth of 1500 km, with spatial resolutions of 250 and 1000 m. Contributes also to atmospheric (cloud properties) and terrestrial (vegetation) applications. Synergism with RA-2, SCATT and AATSR;

MIPAS   Michelson Interferometer for Passive Atmospheric Sounding.
Limb viewing, mid-infrared interferometer for measuring the emission of trace gas species (including ozone mapping and NO_y), temperature, aerosol particles, and polar stratospheric clouds from the upper troposphere up to the mesosphere (8-100 km) with a vertical resolution of 3 km and a horizontal field of view of 30 km at the Earth's limb. Synergism with GOMOS and SCIAMACHY for complementary measurements;

RA-2   Radar Altimeter-2
Adaptive pulse-limited altimeter that measures power level and time position of the earliest part of its radar echoes from ocean, ice and land surfaces and contributes to significant wave height and sea level determination, ocean circulation (ocean dynamics), ice sheet topography and land mapping. Synergism with SCATT, MERIS and AATSR.

The SAR and SCATT use the same frequency band and might be combined in a single instrumentation following the ERS-1/2 technical concept.

Announcement of Opportunity instruments

As a result of an Announcement of Opportunity (AO) for instrument proposals for inclusion in the first polar orbit mission, a set of instruments were selected for possible inclusion in the first mission. The current list comprises the following:

Positioning

PRAREE   Precise Range and Range Rate Experiment Extended version.
Laser/microwave range and range rate system developed for high precision orbitography with potential to contribute to geodesy, plate tectonics and ocean topography. Derived from the PRARE version flown on the ERS satellites. Synergism with RA-2;

Chemistry and dynamics of the upper atmosphere

GOMOS   Global Ozone Monitoring by Occultation of Stars.
Ultraviolet/visible spectrometer, mainly intended for monitoring of global ozone using stellar occultation, in particular for studies of ozone depletion and the greenhouse effect. GOMOS shall provide stable reference data on global ozone, plus observations of H_2O, NO_2, NO_3,

ClO, OClO, temperature and aerosol, in a vertical range between about 20 and 100 km. Synergism with MIPAS and SCIAMACHY;

SCIAMACHY   Scanning Imaging Absorption Spectrometer for Atmospheric Cartography.
A limb and nadir viewing spectrometer operating from the ultraviolet up to the infrared. It will observe reflected and scattered light from tropospheric and stratospheric constituents, temperature and aerosol, with the unique capability to retrieve vertical distribution of numerous atmospheric trace gas species with a vertical resolution of 1 km between 20 and 100 km altitude range in the limb mode, and a 350 m × 25 km field of view in the nadir mode. Synergism with MIPAS and GOMOS;

Radiation

AATSR   Advanced Along-Track Scanning Radiometer.
Conically scanning infrared radiometer dedicated mainly to the retrieval of sea surface temperature, operating in five spectral bands and with a spatial resolution of 1 km over a 500 km swath width. Contribution to ocean dynamics, radiative interactions, cloud parameters and bidirectional reflectance measurements over land surfaces. Synergism with RA-2, SCATT and MERIS;

CERES   Clouds and the Earth's Radiant Energy System.
Broadband scanning radiometer monitoring the reflected and emitted global radiation flux 'on top of the atmosphere' in three spectral bands over an along-track scanning field of view of $\pm 78°$, and a spatial resolution of roughly 25 km at nadir. Intended for studies of the Earth radiation budget and atmospheric radiation, the climate system and atmospheric energetics. CERES is derived from the former ERBE instrument.

The Synthetic Aperture Radar (SAR), the Radar Altimeter (RA-2), the Wind Scatterometer (SCATT), AATSR and PRAREE will be derivatives of ERS-1 instruments, with enhanced capabilities.

Only a subset of the Announcement of Opportunity instruments may fly. Table 14.1 indicates the parameters that will be observed by the various instruments.

Links between the payload and the objectives of the first polar mission

The specific contributions to be made by the various elements of the payload to the three mission objectives of the first polar platform mission are summarized in the following sections.

The meteorology mission

Particularly important for the operational meteorological mission are the data from the operational meteorological package (VIRSR, IASI, IRTS,

Table 14.1 Cross-correlation matrix between instruments and observation parameters.

Instruments disciplines (Parameters)	VIRSR	AMSU	IRTS	SAR	SCATT	ALT-2	MERIS	MIPAS	GOMOS	CERES	SCIAM-ACHY	PRARE	EEA	ATSR
Atmosphere														
Clouds	×	×	×	–	–	–	×	–	–	–	×	–	–	×
Humidity	×	×	×	–	–	–	–	–	–	–	×	–	–	–
Precipitation	–	×	–	–	×	–	–	–	–	–	–	–	–	–
Radiative fluxes	×	×	×	–	–	–	×	–	–	×	×	–	–	×
Temperature	–	×	×	–	–	–	–	×	–	–	×	–	–	–
Trace gases and aerosols	–	–	–	–	–	–	×	×	×	–	–	–	–	–
Wind direction and speed	–	–	–	–	×	–	–	–	–	–	–	–	–	–
Land														
Surface temperature	×	–	–	–	–	–	–	–	–	–	–	–	–	×
Vegetation characteristics	–	–	–	×	–	–	×	–	–	–	–	–	–	–
Surface elevation	×	–	–	×	–	×	–	–	–	–	–	–	–	–
Ocean														
Ocean colour	–	–	–	–	–	–	×	–	–	–	–	–	–	–
Sea surface temperature	×	–	–	–	–	–	–	–	–	–	–	–	–	×
Surface topography	–	–	–	×	–	×	–	–	–	–	–	–	–	–
Turbidity	×	–	–	–	–	–	×	–	–	–	–	–	–	–
Wave height	–	–	–	×	×	×	–	–	–	–	–	–	–	–
Wave spectra	–	–	–	×	×	–	–	–	–	–	–	–	–	–
Wind field	–	–	–	×	×	×	–	–	–	–	–	–	–	–
Marine geoid	–	–	–	–	–	×	–	–	–	–	–	–	–	–
Ice														
Extent	×	–	–	×	–	–	×	–	–	–	–	–	–	×
Snow cover	×	–	–	×	–	–	×	–	–	–	–	–	–	×
Topography	–	–	–	×	–	×	–	–	–	–	–	–	–	–
Temperature	×	–	–	–	–	–	–	–	–	–	–	–	–	×

and AMSU). At the ocean surface many important variables will be observed; key instruments are the wind scatterometer (SCATT), for the observation of wind speed and direction, the radar altimeter (RA-2), for wave height measurements and wind speed, and the AATSR for high-accuracy sea surface temperature measurements.

The ocean mission

Within this mission the most important observables are surface topography, surface wind and wave fields, ocean colour and sea surface temperature; they will be measured, consolidating and extending data sets acquired by the ERS satellites. The key instruments will be the Synthetic Aperture Radar (SAR), the Wind Scatterometer (SCATT), MERIS, AATSR and the Radar Altimeter (coupled with the PRAREE). Data from the elements of the operational meteorological package will also be important.

The climate mission

Key climate monitoring variables span all the elements of the Earth system; in addition to atmospheric variables, observations of surface characteristics (e.g. surface temperatures and land usage) are vital inputs to climate models. Of particular importance is the Earth's radiation balance and here, data from CERES, again supported by data from the operational meteorological instruments, will prove particularly useful. On the polar platform the broad band radiation data will be supported by observations of clouds, aerosols, etc., plus profiles of the two key thermodynamic variables, namely temperature and humidity.

Coupled with these data will be observations of key trace gas species, such as ozone. For studies of trace gas species, the nadir viewing and limb viewing chemical sensors will enable many of the key chemical species to be observed including the complete NO_y family. Observations will be possible in both the troposphere and stratosphere where key instruments are the three chemistry instruments GOMOS, MIPAS and SCIAMACHY. For the Earth's surface, data from AATSR, AVHRR and MERIS will be essential, as will those from the operational meteorological suite of instruments.

The ground segment

The polar satellite can be regarded as a successor to the ERS satellite so Earth observation users will naturally expect a continuation of the services provided by these satellites. They will want to capitalize on their investments in computer hardware and software, as well as the experience gained from previous operations. Discontinuities of service would be most unwelcome as timeliness of data delivery will be essential if the mission is to meet its objectives.

Within this context, the main requirements for the ground segment are compatibility with ERS, the capability to handle the data from the polar platform instruments in a timely fashion and growth potential, so that new possibilities for data handling can be fully exploited. The user interface must therefore ensure that new national facilities can be easily incorporated into the ground segment and that they are designed to meet individual needs. There is also a clear need for an extensive archiving facility.

The main elements of the ground segment structure will include:

(1) a user interface for Earth observation users which increases the mutual user–system understanding;
(2) a central coordination facility which ensures the realization of the Earth observation mission objectives;
(3) receiver stations which acquire data either directly from the polar platform or via a data relay satellite;
(4) a processing delivery and archiving (decentralized) facility which elaborates, distributes and archives the required end data products.

Conclusions

The ESA Earth Observation Programme is conceived within an integrated overall strategy intended to address the problems of concern to mankind and to contribute to a better understanding and monitoring of the Earth's system.

The systems required to realize these objectives are expensive, and no nation can really afford to develop and operate complex space systems and the necessary ground infrastructure. The Agency's Earth Observation Programme accepts this by seeking collaboration, and taking full account of other Earth observation programmes. It is wide ranging being intended to ensure that Europe is able to capitalize properly on the potential of remote sensing from space as well as being in a position to assess the options properly when addressing fundamental problems facing mankind.

Chapter 15

Temperature-independent thermal infrared spectral indices and land surface temperature determined from space

F. BECKER and Z.-L. LI

Groupement Scientifique de Teledetection Spatiale (GSTS),
LSIT[1]/ENSPS,
Strasbourg

Introduction

Remote sensing in the thermal infrared (TIR) bands provides a means of measuring both surface emissivity and surface temperature. Measurements of these two parameters are very useful in many applications, particularly those related to climate studies, for the following reasons:

(1) surface temperature is not only a good indicator of both the energy equilibrium of the Earth's surface and greenhouse effects, but is also one of the key variables controlling fundamental biospheric and geospheric interactions between the Earth's surface and its atmosphere; and

(2) surface spectral emissivities are very useful indicators of specific spectral properties linked with the reststrahlen bands (spectral bands in which there is a broad minimum of emissivity resulting from interatonic stretching vibrations of silicon and oxygen bound in the crystal lattice), and they can lead to the accurate measurement of surface temperature (remember that the error on surface temperature is about fifty times the relative error on the emissivity at a surface temperature of 300 K).

However, besides the effects of the atmosphere, which occur in all the spectral bands, there are several problems, specific to TIR bands, that need to be solved before a relevant quantitative interpretation of the data can be performed. Firstly, as it is well known, the radiance, R_i, measured from space depends on both spectral emissivity and surface temperature. Therefore it is generally not possible to interpret a variation of radiance in the TIR bands in terms either of a temperature variation or of an emissivity variation, as the separation of these two contributions is a complicated task. Secondly, the strong vertical and horizontal heterogeneity of the land surface, at the scale of a pixel, makes it difficult to define effective temperatures and

---

[1]Laboratory associated to CNRS.

emissivities, as has been shown by Becker (1981) and Caselles and Sobrino (1989). In these cases the surface temperature T_s can be defined physically as:

$$T_s = B_i^{-1} \left\{ \frac{R_i - (1 - \varepsilon_i) R_{ati\downarrow}}{\varepsilon_i} \right\}$$

where B_i^{-1} is the inverse of the Planck function in channel i, $R_{ati\downarrow}$ is the downwelling atmospheric radiance, and ε_i the emissivity of an heterogeneous pixel in channel i. This definition of surface temperature is used in the following discussion.

This paper presents a method for separating spectral and thermal effects from the radiance, leading to direct measurement from space of both spectral emissivity and land surface temperature. In the first part of the paper, Temperature-Independent Spectral Indices in TIR (TISI) are defined, and some of their properties are given. Particular attention is given to their complementarity with the NDVI. In the second part of the paper, it is shown how these indices may be used to obtain the land surface temperature and emissivity from AVHRR data only.

Temperature-independent spectral indices in thermal infrared bands

Definition of TISI

Approximation of channel radiance measured at ground level

As given in Becker and Li (1990a), the channel radiance, R_i, measured in channel i at ground level may be written as

$$R_i = B_i(T_{gi}) = \varepsilon_i B_i(T_s) + (1 - \varepsilon_i) R_{ati\downarrow} \tag{1}$$

In this expression, T_{gi} is the surface brightness temperature in channel i at ground level, and $B_i(T)$ is the channel radiance which are closely approximated (Slater, 1980) by the expression

$$B_i(T) = \alpha_i T^{n_i} \tag{2}$$

where α_i and n_i are channel constants for a given reference temperature. Using approximation (2), it is possible to rewrite Equation (1) as

$$R_i = \alpha_i T_{gi}^{n_i} = \varepsilon_i \alpha_i T_s^{n_i} + (1 - \varepsilon_i) R_{ati\downarrow} \tag{3}$$

Definition of TISI

If, for each channel radiance R_i, the atmospheric reflected radiance is neglected, equation (3) becomes

$$T_{gi}^{n_i} = \varepsilon_i T_s^{n_i}$$

By taking the product of N channels with powers $a_k (k=1, \ldots N)$ we obtain

$$\prod_{k=1}^{N} T_{gk}^{a_k n_k} = T_s^P \prod_{k=1}^{N} \varepsilon_k^{a_k} \tag{4}$$

with

$$P = \sum_{k=1}^{N} a_k n_k$$

and choosing the a_k's such that

$$P = 0$$

the surface temperature T_s is eliminated in Equation (4) which yields

$$\prod_{k=1}^{N} T_{gk}^{a_k n_2} = \prod_{k=1}^{N} \varepsilon_k^{a_k} \tag{6}$$

If the atmospheric reflected radiance is taken into account, Equation (3) becomes:

$$T_{gi}^{n_i} = \varepsilon_i T_s^{n_i} C_i$$

where

$$C_i = 1 + \beta_i \frac{1 - \varepsilon_i}{\varepsilon_i}$$

with

$$\beta_i = \frac{R_{ati\downarrow}}{B_i(T_s)}$$

we obtain
with

$$\prod_{k=1}^{N} T_{gi}^{a_k n_k} = C \prod_{k=1}^{N} \varepsilon_k^{a_k} \tag{7}$$

$$C = \prod_{k=1}^{N} C_k^{a_k}$$

As shown in Figure 15.1, C is almost independent of surface temperature T_s and $C \cong 1$ with a good approximation. We define:

$$\text{TISI} = \prod_{k=1}^{N} T_{gk}^{a_k n_k} \tag{8a}$$

$$= M \prod_{k=1}^{N} R_k^{a_k} \tag{8b}$$

where M is a constant given by

$$M = \prod_{k=1}^{N} B_k(T_s)^{-a_k} \cong \prod_{k=1}^{N} \alpha_k^{-a_k}$$

Therefore, if atmospheric reflected radiance is neglected, we obtain from Equation (6)

$$\text{TISI} = \text{TISIE} = \prod_{k=1}^{N} \varepsilon_k^{a_k} \tag{9}$$

and if it is not neglected, we obtain from Equation (7)

$$\text{TISI} = C \times \text{TISIE} \tag{10}$$

with $C \cong 1$.

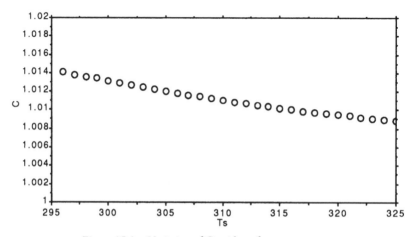

Figure 15.1   Variation of C with surface temperature.

Example of TISI for TIMS data

TIMS is an airborne Thermal Infrared Multispectral Scanner system covering the 8-12 μm region in six spectral bands with a width varying from 0.4–1.0 μm given in Table 15.1 (Palluconi and Meeks, 1985)

Choice of channels

From experimental studies in the laboratory (Nerry *et al.*, 1990) and from the values of the relative emissivities obtained from TIMS data in the Lubbon site as illustrated in Figure 15.2 (Becker and Li, 1990a; Li and Becker, 1990b), it turns out that Channel 3 (2 or 1) is very important for the discrimination of soils and vegetation.

Particular TISI for TIMS

Two particular TISI derived from Equations (8a) and (7) with Channels 3, 4 and 5 of TIMS are given by

$$\text{TISI}_1 = \frac{T_{g3}^{n_3} T_{g5}^{n_5}}{T_{g4}^{2n_4}} = \frac{\varepsilon_3 \varepsilon_5}{\varepsilon_4^2} C_{354} \tag{11}$$

$$\text{TISI}_2 = \frac{T_{g4}^{n_4} T_{g5}^{n_5}}{T_{g3}^{1.8n_3}} = \frac{\varepsilon_4 \varepsilon_5}{\varepsilon_3^{1.8}} C_{453} \tag{12}$$

with $n_3 = 5.162$, $n_4 = 4.816$ and $n_5 = 4.474$.

It should be noted that the powers of ε_3 and ε_4 are different in both cases. The effects of Channel 4 are amplified in TISI_1 while those of Channel 3 are amplified in TISI_2.

Properties of TISI

Equation (10) demonstrates that, within a large range of emissivities and surface temperature, TISI is almost independent of T_s because $C \cong 1$. It is,

Table 15.1   *Spectral bands of TIMS.*

Channels	Bandwidth
1	8.2–8.6
2	8.6–9.0
3	9.0–9.4
4	9.6–10.2
5	10.3–11.1
6	11.3–11.7

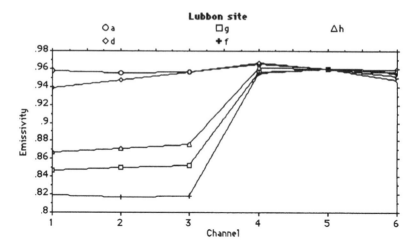

Figure 15.2   The variation of surface relative emissivities obtained from TIMS data when $\varepsilon_5 = 0.96$: g, h and f for bare soil surface samples; a and d for vegetation and forest samples.

therefore, a nearly pure indicator of the spectral properties of the surface since

$$TISI = \prod_{k=1}^{N} \varepsilon_k^{a_k}$$

As emphasized by Becker and Li (1990a), since there are N parameters a_k to determine with only one equation [Equation (5)], it is possible to define several TISI using the same spectral channels in order to weight interesting spectral bands more heavily than others [see Equations (11) and (12)]. TISI are very easy to obtain directly from space since atmospheric effects for TISI are more easily taken into account than for each channel surface brightness as it is shown in Table 15.2. This result shows that TISI can be derived with the

Table 15.2   Illustration of atmospheric effects on TISI: $TISI_b$, obtained with the brightness temperature T_i; $TISI_r$ and $TISI_s$, TISI obtained with T_{gi} derived using actual radiosoundings and the atmospheric standard model.

TISIE				0.941				
Angle			0°				40°	
T_s(K)	290	300	310	320	290	300	310	320
$TISI_b$	0.984	0.993	1.001	1.008	0.994	1.005	1.014	1.022
$TISI_s$	0.942	0.943	0.944	0.945	0.942	0.944	0.945	0.946
$TISI_r$	0.941	0.941	0.941	0.941	0.941	0.941	0.941	0.941

standard atmosphere or with the atmospheric parameters retrieved from satellite soundings if radiosoundings are not available.

Correlation between TISI and other indices

Correlation between TISI and NDVI over HAPEX

Since TISI and NDVI are both connected to surface characteristics in the different spectral region, it is worth analysing correlations between them. During HAPEX (Hydrologic and Atmospheric Pilot EXperiment) (André *et al.*, 1986) the radiometer TIMS was flown simultaneously with the NS001 scanner, and data were acquired over the Lubbon and Castelnau test sites located in the 100×100 km^2 square of HAPEX in southwestern France. From those data, TISI$_1$ was obtained using Equation (11) and NDVI was calculated using the data in Channel 3 (red, 0.633–$0.697 \,\mu$m) and Channel 4 (near-infrared, 0.767–$0.910 \,\mu$m) of the NS001 scanner. Figure 15.3 shows an image of TISI$_1$ and Figure 15.4 shows the correlations between TISI$_1$ and NDVI for the Lubbon test site.

Figure 15.3   *TISI$_1$ image for Lubbon site (TISI$_1$ = NC/100).*

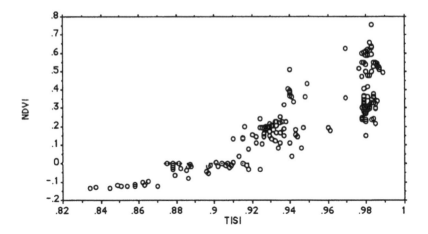

Figure 15.4   Correlation between $TISI_1$ and NDVI for Lubbon site.

Since, in our study site, $\varepsilon_4 \cong \varepsilon_5$, $TISI_1$ may be written as:

$$TISI_1 = \frac{\varepsilon_3 \varepsilon_5}{\varepsilon_4^2} \cong \frac{\varepsilon_3}{\varepsilon_4}$$

$$NDVI = \frac{A_4 - A_3}{A_4 + A_3} \cong \frac{A_4/A_3 - 1}{A_4/A_3 + 1}$$

where A_3 and A_4 are the radiances measured respectively in Channel 3 and Channel 4 of NS001. As shown in Figure 15.2, in our study site, for bare soil $\varepsilon_3 < \varepsilon_4$ and $A_2 \cong A_1$ leading to $TISI_1 \ll 1$ and $NDVI \ll 1$, while, for vegetation, $\varepsilon_3 \cong \varepsilon_4$ and $A_2 \ll A_1$ leading to $TISI1 \cong 1$ and $NDVI \cong 1$. From this figure, the following remarks can be made:

(1)   TISI is more sensitive to bare soil characteristics than NDVI which is, in turn, more sensitive to vegetation. This shows that TISI and NDVI are complementary;
(2)   the particular situations corresponding to points located outside the average curve of those correlations may correspond to pixels having particular states (humidity, nature of surface), as seems indicated be a preliminary study using microwave radiometry (Schmugge *et al.*, 1991). Thus a particular orthogonal index may be constructed with the orthogonal distance to this curve.

Correlation between TISI and surface temperature T_s

As in the visible and near-infrared regions, it is interesting to analyse the correlations between the TISI and the surface temperature. Since, in our study site, the emissivities in Channel 5 are almost spatially constant and are the largest, it can be shown (Li and Becker, 1990) that the surface brightness temperature, T_{g5}, in Channel 5 has the same spatial structure as T_s, and may be used to analyse statistical distributions within an image. Therefore, two bidimensional histograms displaying respectively T_{g5} versus $TISI_1$ (Figure 15.5) and T_s versus $TISI_1$ should be quite similar.

As expected, this histogram shows that:

(1) pixels having the same TISI may show large surface temperature differences as shown in Figure 15.6. TISI is, therefore, a good representation of the thermal heterogeneity of a given medium and a good indicator of the impact of surface characteristics on the thermal heterogeneity;

(2) surfaces having the same temperature may have different TISI as shown in Figure 15.7. This shows that different surfaces (i.e. different TISI) may have the same surface temperature T_s.

(3) considering the average behaviour of $TISI(T_s)$ for the data acquired at noon, we see that, in this example, the pixels with high T_s and low $TISI_1$ correspond to bare soils, while the pixels with low T_s and high $TISI_1$ correspond to vegetation.

Determination of emissivities and land surface temperature from AVHRR data

Principle of the method

It is well known that in Channel 3 of AVHRR, i.e. around 3.7 μm, the radiance emitted by the surface itself and the reflected radiance due to sun

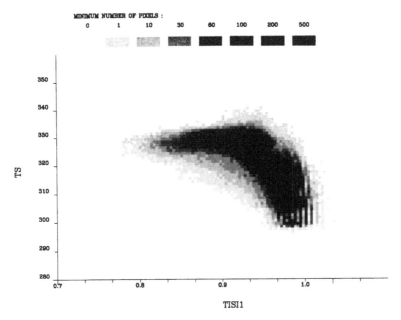

Figure 15.5 Illustration of correlations between $TISI_1$ and surface temperature (T_{g5}).

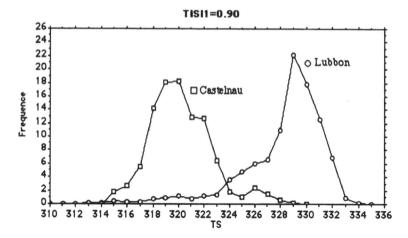

Figure 15.6   *Histogram of surface temperature for* $TISI_1 = 0.90$.

irradiation during the day, are of the same order of magnitude. Using this
very interesting property, Becker and Li (1990a) proposed a method for
determining the surface temperature and surface emissivities from AVHRR
using only the sun as an active source. This is displayed on Figure 15.8 which
shows the difference of radiance measured at ground level between the day
and the night in Channel 3 of AVHRR (3.7 μm).

The general idea behind this method is to use the radiation emitted during
the night to evaluate the radiation emitted during the day, thanks to the non-
dependence of TISI on T_s, which takes care of the fact that the surface
temperature T_s during day and night are very different.

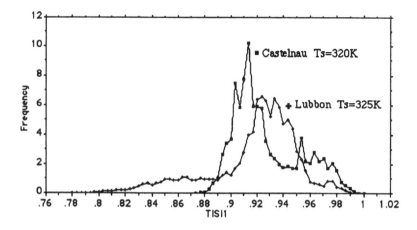

Figure 15.7   *Histogram of* $TISI_1$ *for a given surface temperature.*

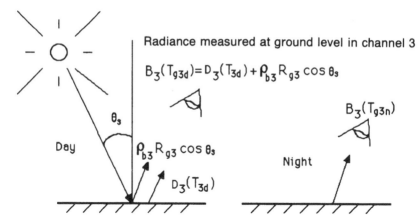

Radiance measured at ground level in channel 3

$$B_3(T_{g3d}) = D_3(T_{3d}) + P_{b3}R_{g3}\cos\theta_s$$

$$B_3(T_{g3n})$$

Day

$$P_{b3}R_{g3}\cos\theta_s$$

$$D_3(T_{3d})$$

Night

Figure 15.8   Difference of radiance measured at ground level between the day and the night in Channel 3 (3.7 μm) of AVHRR.

From the definition of TISI [Equation (8b)], it is possible to construct a TISI with 3 channels of AVHRR given by

$$\text{TISI} = M \frac{B_3(T_{g3})}{B_4(T_{g4})^{a_4} B_5(T_{g5})} \qquad (13)$$

with $a_4 = 1.924$ and $M = 0.62105 \times 10^{-7}$ for NOAA 11.

During the night, since there is no solar reflection, it is shown from Equations (13), (8b) and (9) that

$$\text{TISI}_n = M \frac{B_3(T_{g3n})}{B_4(T_{g4n})^{a_4} B_5(T_{g5n})}$$

$$\cong \text{TISIE}_n = \frac{\varepsilon_3}{\varepsilon_4^{a_4} \varepsilon_5} \qquad (14)$$

with a very good approximation, where the emissivities take on their night values.

During the day, solar reflection is important in Channel 3 of AVHRR and the radiance $B_3(T_{g3d})$ measured at ground level in this channel is given by

$$B_3(T_{g3d}) = D_3(T_{3d}) + \rho_{b3}(\theta_s,\, \theta) \times R_{g3}(\theta_s) \cos \theta_s \qquad (15)$$

where $D_3(T_{3d})$ is the channel radiance which would have been measured in Channel 3 if there were no solar reflection, $\rho_{b3}(\theta_s,\, \theta)$ is the bidirectional reflectivity of the surface in Channel 3 in the direction of θ_s, θ, while $R_{g3}(\theta_s)$ is the solar irradiance at the surface in Channel 3, and θ_s, θ are respectively the angle of solar irradiation and the angle of observation.

Therefore, as was done for the night-time data, a good approximation is:

$$M\frac{D_3(T_{3d})}{B_4(T_{g4d})^{a4}B_5(T_{g5d})} \cong TISIE_d = \frac{\varepsilon_3}{\varepsilon_4^{a4}\varepsilon_5} \qquad (16)$$

where the emissivities take on their day values.

Inserting expressions (15) and (16) into the definition of $TISI_d$ [Equation (13)], we get:

$$TISI_d = TISIE_d + M\frac{\rho_{b3}(\theta_s, \theta) \times R_{g3}(\theta_s) \cos \theta_s}{B_4(T_{g4d})^{a4}B_5(T_{g5d})} \qquad (17)$$

It is assumed that emissivity ratios do not vary substantially between day and night, so that $TISIE_d = TISIE_n$. This assumption has been checked and shown to be reasonable for the particular data set measured *in situ* (Labed, 1990) and shown in Table 15.3 for $TISIE_{45}$.

Using expressions (14) and (17) one can obtain the bidirectional reflectivity $\rho_{b3}(\theta_s, \theta)$ in Channel 3:

$$\rho_{b3}(\theta_s, \theta) = \frac{(TISI_d - TISI_n)B_4(T_{g4d})^{a4}B_5(T_{g5d})}{MR_{g3}(\theta s) \cos \theta_s}$$

If one writes, for opaque media, the bidirectional reflectivity of the surface in Channel 3, $\rho_{b3}(\theta_s, \theta)$, as

$$\rho_{b3}(\theta_s, \theta) = \frac{1 - \varepsilon_3(\theta)}{\pi}f_3(\theta_s, \theta)$$

where $f_3(\theta_s, \theta)$ is the angular form factor of the surface (which is unity if the surface is a Lambertian reflector), one finally obtains

$$\varepsilon_3(\theta) = 1 - \frac{\pi(TISI_d - TISI_n)B_4(T_{g4d})^{a4}B_5(T_{g5d})}{MR_{g3}(\theta_s) \cos \theta_s f_3(\theta_s, \theta)} \qquad (18)$$

Then, from the definition and properties of TISI, it is very easy to obtain the emissivities in Channels 4 and 5 using the following formulae:

Table 15.3 Comparison of $TISIE_{45}$ ($TISIE_{45} = \varepsilon_4/\varepsilon_5^{1.095}$) between day and night using the emissivities measured by Labed (1990).

Parameters	In situ measurements	
	Day	Night
ε_4	0.978	0.957
ε_5	0.994	0.976
$TISIE_{45}$	0.984	0.983

$$\varepsilon_4 = \left\{ \frac{\varepsilon_3}{TISI_{54} \times TISI_n} \right\}^{1/(n_{54}+a_4)} \tag{19}$$

$$\varepsilon_5 = \varepsilon_4^{n_{54}} TISI_{54} \tag{20}$$

with $n_{54} = 0.913$.

Expressions (18), (19) and (20) show that the emissivities can be retrieved directly from space, provided that $f_3(\theta_s, \theta)$ and $R_{g3}^s(\theta_s)$ are known. This solves the first part of the problem.

Once the emissivities in Channels 4 and 5 are known, the surface temperature is then easily derived from the brightness temperatures T_4 and T_5 measured by AVHRR using for instance the local split window method (Becker and Li, 1990b) which reads

$$T_s = A_0 + P \frac{T_4 + T_5}{2} + M \frac{T_4 - T_5}{2}$$

where

$$A_0 = 1.274$$

$$P = 1 + 0.15616 \frac{1-\varepsilon}{\varepsilon} - 0.482 \frac{\Delta\varepsilon}{\varepsilon^2}$$

$$M = 6.26 + 3.98 \frac{1-\varepsilon}{\varepsilon} + 38.33 \frac{\Delta\varepsilon}{\varepsilon^2}$$

$$\varepsilon = \frac{\varepsilon_4 + \varepsilon_5}{2}, \quad \Delta\varepsilon = \varepsilon_4 - \varepsilon_5$$

It should be noted that the method which we propose here has some advantages over other approaches. First and foremost, it allows retrieval both the spectral emissivities and surface temperature from space using only the passive radiometry. Secondly, provided that the surface characteristics do not change, it is not necessary to use successive day and night images because TISIs are independent of surface temperature. The third advantage of this method for determining emissivity is a strong reduction of the impact of errors of measurement. Since the relative accuracy of the emissivity is related to the relative accuracy of the reflectivity by:

$$\frac{\Delta\varepsilon_3}{\varepsilon_3} = \frac{1-\varepsilon_3}{\varepsilon_3} \times \frac{\Delta\rho_{h3}}{\rho_{h3}}$$

The relative error on ε_3 may be much smaller than the relative error on the reflectivity ρ_{h3} by a factor 5 for $\varepsilon_3 = 0.75$ to 50 for $\varepsilon_3 = 0.98$. However, this method has two important drawbacks (which appear also in the determina-

VALEUR DE DEBUT DE CLASSE :

| 74 | 85 | 89 | 90 | 91 | 92 | 93 | 94 |

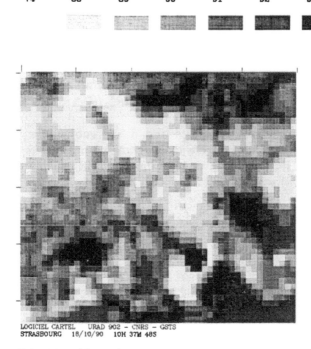

Figure 15.9   Map of day surface emissivity in Channel 3 over La Crau for 19 June 1989. The grey scale ranges from 0.84 (bright) to 0.96 (dark).

tion of the spectral albedo). It needs a knowledge of both the angular form factor, $f_3(\theta_s, \theta)$ and the solar irradiance at the surface in Channel 3, $R_{g3}^s(\theta_s)$.

Preliminary results

Two NOAA-11 AVHRR images acquired respectively on 19 June 1989 (day image) and 23 June 1989 (night image) over the La Crau site in southern France are used here to test the proposed method. Since it was not possible to acquire two clear images corresponding to two consecutive days, the application proposed here is therefore based on the hypothesis that the TISIE are the same during the period 19–23 June 1989, which is a realistic assumption as the surface is a composed of a mixture of stones and dry soils and no rain was observed during that period of time.

Applying this method to the AVHRR data described above, we obtained the emissivities in Channels 3, 4 and 5 for day. Figure 15.9 shows the image of emissivity obtained in Channel 3 while Figure 15.10 displays the histograms of those emissivities in Channels 3, 4 and 5. From their histograms, we notice that in the region of 50×60 km² over the whole image of La Crau, the emissivity in Channel 3 varies mainly from 0.86 to 0.96, and the emissivities in Channels 4 and 5 vary mainly from 0.97 to 1.00. This is substantially in agreement with the measurements of Nerry *et al.* (1990), which show very small variations of emissivities in Channels 4 and 5 of AVHRR, but much larger variation in Channel 3.

Since we have no *in situ* synchronized measurements of day surface temperature in Crau, but only *in situ* measurements of night surface temperature, we calculated the night surface temperature using the retrieved surface day spectral emissivities. In order to compare the results obtained by this method with the *in situ* synchronized measurements (Labed, 1990), the average of the night surface temperatures and day surface emissivities over the zone of about 10×10 km² pixels centred on La Crau and their associate rms are calculated and shown in Table 15.4.

From this table we can make some remarks:

(1) the dispersion is very small, thus confirming the thermal homogeneity of La Crau at the scale of an AVHRR pixel,

(2) the surface temperature and emissivities (ε_4, ε_5) obtained from AVHRR data are in good agreement with *in situ* measurements. This shows the feasibility of our method, and

(3) the emissivity in Channel 3 obtained from AVHRR is different from the *in situ* measurement. This may be due to the fact that the *in situ* measurement of this emissivity (Channel 3) is not a direct emissivity measurement. As pointed out by Labed (1990), this value of emissivity is only indicative because its precision is very poor, of the order of 10 per cent.

Conclusions

This work has shown that:

(1) spectral analysis in TIR bands can be performed using TISI which are independent

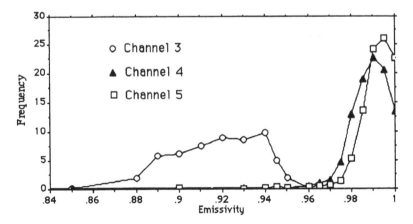

Figure 15.10   Histograms of day surface emissivities in Channels 3, 4 and 5 of AVHRR.

Table 15.4   Comparison between satellite measurements and the ground measurements.

Parameters	Measurements from AVHRR		In situ measurements	
	Day	Night	Day	Night
$\varepsilon_3 + \sigma$	0.890 ± 0.003	—	0.830†	0.830†
$\varepsilon_4 + \sigma$	0.981 ± 0.002	—	0.978	0.957
$\varepsilon_5 + \sigma$	0.989 ± 0.002	—	0.994	0.976
$T_s + \sigma$	—	$*19.6 \pm 0.15$	—	19.8

†This value is not derived from the direct emissivity measurement, it is thus only the indicative value.

*This value is obtained using the retrieved surface day spectral emissivities (column 2) from AVHRR data. See text.

of surface temperature and are very easy to derive directly from space. In the examples discussed, it appears that TISI and NDVI are complementary and can lead to stronger results if used together rather than separately. This is confirmed by the fact that the behaviour of $TISI(T_s)$ is similar to that of $NDVI(T_s)$,

(2) both the land surface temperature and channel emissivities can be derived from the radiances measured by AVHRR using TISI. The results obtained using actual AVHRR data acquired over La Crau test site by assuming that the surface is a Lambertian reflector turn out to be quite satisfactory, and

(3) for future space radiometry, it appears that for interpretation of thermal infrared data it is very useful to use a radiometer with at least four channels, namely one channel in the 3.7 μm band (day and night), a second channel in the 8-9 μm band (before the ozone band) and two channels in the window 10-12 μm for the interpretations of the thermal infrared data.

Acknowledgments

We are grateful to Professor M.P. Stoll, Director of GRTR/LSIT, for valuable discussions, and to Professor Paul M. Mather, Chairman of The Remote Sensing Society, for helpful reviews and editorial changes in the manuscript.

References

André, J.C., Goutorbe, J.P. and Perrier, A., 1986, HAPEX-MOBILHY: A hydrologic atmospheric experiment for the study of water budget and evaporation flux at climatic scale, Bulletin of the American Meteorological Society, **67**, 134-44.

Becker, F., 1981, Angular reflectivity and emissivity of natural media in the thermal infrared bands, in Guyot, G. et al. (Eds), First International Colloquium on Spectral Signatures of Objects in Remote Sensing, Avignon, 57-72. INRA.

Becker, F. and Li, Z. L., 1990a, Temperature-independent spectral indices in thermal infrared bands, Remote Sensing of Environment, **32**, 17-33.

Becker, F. and Li, Z. L., 1990b, Towards a local split window method over land surface, *International Journal of Remote Sensing*, **3**, 369–93.

Caselles, V. and Sobrino, J.A., 1989, Determination of frosts in orange groves from NOAA-9 AVHRR Data, *Remote Sensing of Environment*, **29**, 135–46.

Labed, J., 1990, *Approche expérimentale des problèmes liés à l'émissivité et à la température de surface, ainsi qu'à leur variabilité spatio-temporelle, dans le cadre de la télédétection spatiale dans l'infrarouge thermique*, PhD thesis, University of Strasbourg. September, 1990.

Li, Z.L. and Becker, F., 1990, Properties and comparison of temperature-independent thermal infrared spectral indices with NDVI for HAPEX data, *Remote Sensing of Environment*, **33**, 165–82.

Nerry, F., Labed, J. and Stoll, M.P., 1990, Spectral properties of land surfaces in the thermal infrared band. Part I: Laboratory measurements of absolute spectral emissivity and reflectivity signatures, *Journal of Geophysical Research*, **95**, 7027–44.

Palluconi, F.D. and Meeks, G.R., 1985, *Thermal Infrared Multispectral Scanner TIMS, An investigator's guide to TIMS data*, NASA, Jet Propulsion Laboratory, Pasadena, California.

Schmugge T.J., Jackson, T.J. and Wang, J.R., 1991, Passive microwave remote sensing of soil moisture: results from HAPEX, FIFE and MONSOON 90, *5th International Colloquium on Physical Measurements and Signatures in Remote Sensing*, Courchevel, France, ESA-SP319, pp. 315–19.

Slater, P.N., 1980, *Remote Sensing Optics and Optical Systems*, Reading, MA: Addison-Wesley, 246–7.

Chapter 16
Extracting surface properties from satellite data in the visible and near-infrared wavelengths

M.M. VERSTRAETE[1] and B. PINTY[2]

[1] *Institute for Remote Sensing Applications,*
Joint Research Centre of the CEC,
Ispra, Italy

[2] *Laboratoire d'Etudes et de Recherches en Télédétection Spatiale,*
Toulouse, France

Abstract

The principles of satellite remote sensing, based on the measurement of electromagnetic radiation and the process of inversion of these data, are briefly reviewed. The implications for the retrieval of quantitative information on the nature, structure and evolution of surfaces on Earth are discussed, and a coherent strategy for the acquisition of biological or other relevant information on these environments is outlined.

Introduction

The TERRA-1 conference was convened explicitly to assess the role of space-borne observations in monitoring environmental processes on Earth. The rationale for studying environmental degradation on a global basis, and from an interdisciplinary perspective, has been extensively discussed in a number of papers on the Global Change Research Programme, the International Geosphere-Biosphere Programme, and similar documents describing national and international research activities (for example, IGBP, 1988; NAS, 1988; CNES, 1989). Up-to-date overviews of these issues and programmes can be found in this volume.

Satellite remote sensing is the only technique that allows the systematic monitoring, on a global and repetitive basis, of the atmospheric, oceanic, terrestrial and biological components of the climate system, at the space- and timescales of interest, and with the required resolution (e.g. NASA, 1987). A detailed discussion of the relative advantages and drawbacks of different observation techniques (including *in situ* observations and measurements based on the remote sensing of electromagnetic radiation) is outside the scope of this paper. However, some of the issues relating to the scale of representativity of each type of measurement have been discussed by Verstraete and Pinty (1991). The purpose of this paper is to identify some of the issues associated with the retrieval of quantitative information from satellite remote sensing measurements in the visible and near-infrared spectral regions. The role of physically-based models is underlined, and some of the requirements of a useful inversion procedure are outlined.

The nature of remote sensing

For the vast majority of applications, and especially for most space-borne observations of the planet, the term 'remote sensing' implies the measurement of an electromagnetic radiation field. In the case of the visible and near-infrared spectral bands, the radiation, ultimately coming from the sun, interacts with the atmosphere and the surface of the Earth before reaching the sensor on board the satellite. A variety of issues derive from these simple facts and must be addressed. First of all, the characteristics of the radiation measured by the sensor result from all the processes which have controlled its emission and scattering during its lifetime. For the purpose of this paper, we will assume that the satellite measurements have been calibrated in terms of reflectances. It is therefore necessary to assess to what extent the measured signal is representative of the atmospheric or of the surface conditions. Second, since only radiative processes can affect the propagation of light, it must also be clear that the only quantitative information that can possibly be retrieved from satellite remote sensing measurements must be related to these radiative processes. In the rest of this paper, we will focus on the retrieval of information on the surface of the planet and, therefore, restrict ourselves to the spectral regions where the atmosphere is mostly transparent to radiation. The quantitative interpretation of satellite data is an attempt to understand which physical variables and processes have been responsible for producing the radiation that is being measured. For example, since radiation is differentially reflected by different surfaces (soils, vegetation canopies), we may be able to learn something about the reflectance of these surfaces by analysing such data. It follows that it is impossible to obtain any information on processes or variables that do not directly affect the reflectance observed by satellites. Satellite measurements are relative to an area on the ground known as a pixel. These pixels have typical linear dimensions of a few metres to a few kilometres, depending on the sensor type and purpose, and they may contain a number of surfaces or scatterers. The measurement is relative to the entire area seen by the sensor and is, in first approximation, a weighted average of the reflectances of the various surfaces in the pixel. This can be expressed as follows: if (the pixel contains $i = 1, 2, \ldots n$ surfaces, each characterized by a reflectance ρ_i and a fractional area coverage σ_i, then

$$\rho_{\text{pixel}} = R(x_1, x_2, t, \lambda; \rho_i, \sigma_i, X_{ij}, A_k) \tag{1}$$

where the first series of variables (x_1, x_2, t, λ) represents the location, time, and spectral interval of the observation, X_{ij} represents non–linear interaction terms between the different surfaces, and A_k $(k=1,2, \ldots, m)$ stands for the atmospheric parameters that may affect the measurements.

On the basis of our current understanding of the process of radiation transfer in the visible and near-infrared spectral regions, we can further describe the reflectance of homogeneous optically deep surfaces as functions of a single scattering albedo ω, a phase function parameter Θ, a scatterer orientation distribution parameter κ, and one additional parameter S for the macroscopic description of the structure of the medium. This can be formally written as follows:

$$\rho_i = r(\theta_1, \theta_2, \phi, \lambda; \omega_i, \Theta_i, \kappa_i, S_i) \tag{2}$$

where θ_1 and θ_2 are the illumination and viewing zenith angles, respectively, and ϕ is the azimuth of the viewing direction relative to the illumination direction. Of course, additional parameters may have to be added to describe the reflectance of more complex inhomogeneous surfaces.

The process of inversion

If we knew perfectly the amount and spectral distribution of incoming solar radiation as well as the states of the atmosphere and the surfaces with which the radiation interacts, and if we completely understood the nature of these interactions, we could predict the radiation levels or the reflectances that the satellite sensor should measure. In reality, we measure a radiation level with an imperfect instrument, and we would like to know what was the state of the atmosphere and/or the surface at the time of the measurement. This is known as an inverse problem.

In general, the inverse problem can be formulated as follows: let

$$z = f(x_1, x_2, \ldots, x_n; y_1, y_2, \ldots, y_m) \tag{3}$$

be an analytical representation of the relation between the physical parameters $y_j(j=1, \ldots, m)$ and the estimated observable value z, when the conditions of the observation are described by the independent variables $x_i(i=1,2, \ldots, n)$.

Whenever we make a single measurement of the observable variable z, we obtain a value $\hat{z}$, which is presumably resulting from particular values of y_j. Since we usually have $m > 1$ unknowns and only one equation, the values of y_j cannot be determined on the basis of this unique measurement and of the single equation. We can make multiple measurements (say M), but then we get, again in principle, a system of M equations [Equation (4)] in $M \times m$ unknowns where the z_k on the left hand side of these equations represent the theoretical values corresponding to the indicated variables x and parameters y.

$$z_1 = f_1(x_{11}, x_{12}, \ldots, x_{1n}; y_{11}, y_{12}, \ldots, y_{1m})$$

$$z_2 = f_2(x_{21}, x_{22}, \ldots, x_{2n}; y_{21}, y_{22}, \ldots, y_{2m})$$

$$\vdots$$

$$z_M = f_M(x_{M1}, x_{M2}, \ldots, x_{Mn}; y_{M1}, y_{M2}, \ldots, y_{Mm})$$

(4)

This system can only be simplified and solved if we make additional assumptions. Specifically, we will assume that (i) the target does not change significantly between measurements, i.e. that the form of the equation f and the values of the parameters y_j are unchanged for all measurements, (ii) different observations, taken for various values of the independent variables x_i, display significant variability, and (iii) more observations are taken for various conditions x_i than there are parameters y_j to retrieve ($M > m$). Optionally, one or more of the parameters y_j can be specified on the basis of other sources of information, in which case the number of physical parameters y_j to be retrieved by inversion is decreased. When these conditions are met, the system (4) becomes:

$$z_1 = f(x_{11}, x_{12}, \ldots, x_{1n}; y_1, y_2, \ldots, y_m)$$

$$z_2 = f(x_{21}, x_{22}, \ldots, x_{2n}; y_1, y_2, \ldots, y_m)$$

$$\vdots$$

$$z_M = f(x_{M1}, x_{M2}, \ldots, x_{Mn}; y_1, y_2, \ldots, y_m)$$

(5)

In general, there are differences between the actual measurement values $\hat{z}$ and the theoretical values z obtained on the basis of these equations, because there are errors in the measurements, or because the equations do not perfectly represent the reality. The objective, then, is to minimize, in some statistical sense, these differences, while imposing the form of the model f and the uniqueness of the values of the parameters y_j. Mathematically, we search for the values of y_j that minimize the expression

$$\delta^2 = \sum_{k=1}^{M} [\hat{z}_k - f(x_{k1}, x_{k2}, \ldots, x_{kn}; y_1, y_2, \ldots, y_m)]^2$$

(6)

Standard numerical algorithms are available to perform this minimization and produce the optimal values of the parameters y_j that account for the observed variability of the measurements $\hat{z}$. For this approach to work on an operational basis, however, the numerical algorithms must be able to optimize globally, i.e. be insensitive to local minima, the minimum of the function δ^2 must be well-defined for each of the parameters y_j, and the influence of noise in the data on the retrieved values must be fully documented, so that some known degree of confidence can be associated with the retrieved values y_j (Pinty *et al.*, 1989; Pinty and Verstraete, 1991).

Application to remote sensing in the solar spectral region

The implications of the remarks in the preceding sections are that only those parameters that directly enter into the physical equations describing the radiative processes can be retrieved from an analysis of satellite remote sensing data. In the case of observations in the visible or near-infrared spectral region, and for an homogeneous and optically deep medium, those parameters are the single scattering albedo, the phase function of the scatterers, or the geometric structure of the observed surface (e.g. Verstraete *et al.*, 1990; Pinty *et al*, 1990). For other media, or in other spectral regions, different physical parameters would be obtained.

Although these parameters may have an intrinsic interest for some applications, we are usually more concerned about geochemical, biological or economic variables such as the net carbon flux into the biosphere, the primary productivity of agricultural systems, etc. These parameters must be similarly retrieved from the inversion of chemical or biological models linking these variables of interest to the optical and structural properties retrieved from remote sensing. This may in fact require a suite of models, in the case of complex variables.

It appears that such chemical or biological models do not exist, or should be significantly improved. In fact, this lack of quantitative knowledge on the relations between the state of the soils and plants on one side and their optical and structural properties on the other side may be the single most important limitation to the quantitative utilization of satellite remote sensing data in these fields. We therefore suggest that this area of research should receive the highest priority of support in the coming years, as few of the objectives of the IGBP/Global Change programmes will be reached without a good understanding of the relevant systems and an operational capability to retrieve quantitative information from remote sensing data. As hinted in the introduction, satellites provide our sole source of coherent, global and repetitive data on the planet, but the extraction of useful information from these data depends entirely on our ability to model all the processes that link the variables of interest with the radiation reflected by these surfaces.

Most of the theoretical arguments described in these pages are applicable to other spectral bands, but the physical parameters that can be retrieved would be different: they include the skin temperature of the objects in the pixel, in the case of thermal observations, and physical parameters such as the dielectric properties and water content of these surfaces in the case of radar remote sensing. In all cases, physical parameters are retrieved from the first inversion, and successive orders of inversions with appropriate models must be applied to retrieve the variables of interest. Clearly, significant advances should be expected from the combined analysis of remote sensing data of the same surfaces in different spectral bands.

Much of the work that has been done so far with remote sensing data has followed a more empirical approach, based on image processing techniques and statistical relations between satellite measurements and variables of interest measured on the ground. This initial stage was very useful in establishing and promoting the potential benefit of satellite remote sensing. More quantitative approaches must now be developed, however, if we are to address the fundamental issues identified in the introduction, and this requires the detailed modelling of all the processes that link the variables of interest with the radiation signals measured on board satellites.

Acknowledgments

The perspective on remote sensing described in this paper has been acquired and progressively refined over the last couple of years. During this period of time, we have been supported by the USGS EROS Data Center, Sioux Falls, SD, under contract number 14-08-00001-A0723. Further support was provided by the French Programme National de Télédétection Spatiale and the Exploratory Research Programme of the Joint Research Centre. The assistance of all of these institutions is gratefully acknowledged.

References

CNES, 1989, *Global Change*, Paris: Centre National d'Etudes Spatiales.
IGBP, 1988, *Global Change Report No. 4: The International Geosphere-Biosphere Programme: A Study of Global Change*, Stockholm: IGBP Secretariat.
NASA, 1987, *From Pattern to Process: The Strategy of the Earth Observing System*, EOS Science Steering Committee Report, Volume II, Washington, DC: National Aeronautics and Space Administration.
NAS, 1988, *Toward an Understanding of Global Change*, Washington, DC: National Academy Press.
Pinty, B. and Verstraete, M.M., 1991, Extracting information on surface properties from bidirectional reflectance measurements, *Journal of Geophysical Research*, **96**, 2865-74.
Pinty, B., Verstraete, M.M. and Dickinson, R.E., 1989, A physical model for predicting bidirectional reflectances over bare soil, *Remote Sensing of Environment*, **27**, 273-88.
Pinty, B., Verstraete, M.M. and Dickinson, R.E., 1990, A physical model of the

bidirectional reflectance of vegetation canopies. Part 2: Inversion and validation, *Journal of Geophysical Research*, **95**, 11767–75.

Verstraete, M. M. and Pinty, B., 1991, The potential contribution of satellite remote sensing to the understanding of arid lands processes, *Vegetatio*, **91**, 59–72.

Verstraete, M. M., Pinty, B. and Dickinson, R.E., 1990, A physical model of the bidirectional reflectance of vegetation canopies. Part 1: Theory, *Journal of Geophysical Research*, **95**, 11755–65.

Chapter 17
Remote sensing and geographical information systems

P.M. MATHER

Department of Geography,
University of Nottingham

Introduction

In recent years the Earth sciences have witnessed a philosophical shift from a position of exclusivity and exceptionalism towards one of complementarity and synergism. This move has been articulated most clearly by global change researchers who recognize that ' . . . Future research in Earth remote sensing will be directed towards understanding how the planet functions as an integrated system. In the past few decades, we have learned that there are many complex interactions among the various components of the Earth: the geosphere, hydrosphere, cryosphere, atmosphere and biosphere. We have become aware that humans are effecting changes in certain of these components that are of truly global significance. These discoveries have led to an increased recognition that we must develop a truly multidisciplinary approach to the study of how the Earth functions as a whole and of how it might be expected to change on time scales of relevance to humans' (Wickland, 1989).

This change in attitude, towards an emphasis on integrated and interacting Earth systems, has been accompanied by technical and computing developments which permit the storage, retrieval, manipulation and display of spatial information at different scales, both temporal and geographical. Remotely-sensed images are now seen as a particular type of spatial data, with their own spatial and temporal scales, and the incorporation of remotely-sensed data within an integrated spatial data handling system is seen to be the most effective way to extract and use the information they contain.

An integrated spatial data handling, or geographical information system (GIS), provides an organized environment for the execution of a number of related functions, which are described by Rhind and Green (1988) and are summarized under the following headings: data input, encoding and editing; manipulation and analysis; data display and database management. Data input and encoding includes both the capture of data and its validation and

211

editing, as well as structuring, for example by the generation of topology. Manipulation functions are those which relate to the representation of the data, for example geometrical transformations, raster to vector conversion, line generalization, and image enhancement. Data analysis encompasses statistical functions, spatial analysis, and the calculation of derived quantities such as areas and lengths. Database management provides access to the data in a controlled way. The major characteristic of a GIS is that operations on multi-source data are permitted. Multi-source data includes non-spatial (tabular or textual) information as well as spatial data in the form of maps and images.

If they are to be effective tools in Earth and environmental science, GIS must offer, in addition to the functions described by Rhind and Green (1988), the ability to specify and execute models based upon information stored in the database. The development of models to explain, and ultimately predict, the functioning of ecological, hydrological, oceanographic, atmospheric and other environmental systems has always been an important component of Earth science research. However, many of the most significant processes in ecological research, for example, are not amenable to observation by remote sensing. Remotely-sensed data can be used, with other appropriate data, as input to models from which these processes (such as evapotranspiration, photosynthesis, respiration and decomposition) can be estimated indirectly (Peterson and Running, 1989). Examples of models that incorporate remotely-sensed data are Running and Coughlan's (1988) forest ecosystem process model, which requires meteorological and biophysical inputs, derived from satellite observations, and predicts daily rates of evapotranspiration, photosynthesis and respiration. The simple biosphere model of Sellers and Dorman (1987) and the Biosphere–Atmosphere Transfer Scheme of Dickinson (1984) and Wilson *et al.* (1987), are examples of models using remotely-measured parameters (such as leaf area index, canopy height, canopy reflectance and surface roughness). Such models cannot function using remotely-sensed data alone, hence they are best developed within the context of a GIS.

Conventional GIS are not very efficient in their handling of very large spatial databases. The user interface to such databases is often not a simple one; it is partly a computer science problem (how to access large databases quickly and efficiently) and partly a problem in the design of effective user interfaces. The concept of metadata, which is systematic descriptive information about the content, organization and use of the datasets making up a database, is worthy of further exploration. EOSDIS appears to offer at least a partial solution to the problem of how to access data easily. Large-scale databases might include, as part of their stored metadata, various higher-order properties of the data. In the case of remotely-sensed data, such properties would include geometric, atmospheric and other corrections, together with the results of image processing operations, such as the contents of look-up tables.

While it is important to have an informed knowledge of GIS problems in order to make efficient and effective use of the technology, it is perhaps more important to be aware of the difficulties involved in using remotely-sensed image data within a GIS framework. The main section of this paper is devoted to a survey of these difficulties, which provide an impediment to the more widespread use of remotely-sensed data among the terrestrial science community.

Problems in the use of remotely-sensed images
Overview

Remote sensing in the solar and thermal regions of the spectrum is weather-dependent. Cloud cover limits remote sensing of many humid temperate and tropical regions of the world. The advent of sensors with the capability to look off-nadir, helps to alleviate the problem, which is most severe for platforms with a long orbital repeat cycle. LANDSAT, for example, returns regularly every 16 days giving approximately 23 viewing opportunities per year. In some regions, for example the north of Scotland, only a few cloud-free images are serendipitously available over a period of several years. A study by Legg (1991) concludes that '... applications requiring frequent imagery within specific time intervals are impractical in the United Kingdom'.

Microwave sensors are weather-independent and are also capable of operating at night so that a clear image of a target area will be obtained on every overpass. Experience in the interpretation and use of microwave images is, however, not as great as that which has been built up over almost 20 years with images collected in the solar region of the spectrum. As the number of remote sensing platforms of different types increases, and with the advent of EOS, it is expected that a greater number of images of a target area will be acquired for a given time-period. Nevertheless, significant research problems remain. These problems are considered in more detail in the following sections. They relate to the need for sensor calibration, correction for atmospheric, illumination and viewing geometry effects, and georeferencing of imagery. A valuable recent critique is provided by Duggin and Robinove (1990).

Radiometric calibration

Passive imaging sensors, carried by orbiting satellites, collect upwelling radiance from the Earth's surface in one or more spectral bands for each of a large number of areas or pixels on the ground or sea surface. The output of the sensor is a voltage Q which is proportional to the spectral radiance L_λ; the relationship between input and output is approximately linear so that Q is a function of L_λ and the spectral responsivity of the sensor. Asrar (1989) shows that this relationship is, in fact, more complex and that Q is a function of several variables, and that in order to compute the mean radiance for a given spectral band the gain and offset of the sensor must be known as well as the band-pass limits. The gain and offset of the sensor are found initially by pre-launch calibration, and are estimated over the effective lifetime of the instrument by in-flight calibration or by means of measurements made over known targets. Details of pre-launch calibration are normally given in instrument manuals, and in-flight calibration results are stored in the header records of data tapes. The data provided on magnetic tape are integer counts derived from analogue to digital conversion of the voltages output by the sensor. For example, the first three spectral bands of LANDSAT's MSS, which cover the spectral ranges 0.5-0.6 μm, 0.6-0.7 μm and 0.7-0.8 μm respectively, the output voltages are passed through non-linear amplifiers and then quantized on a 6-bit (0-63) scale. These values are transmitted to the ground station, where the counts are linearly decompressed onto a 0-127 scale using look-up tables derived from a knowledge of the characteristics of the non-linear amplifier and the gain and offset of the system (Freden and Gordon, 1983). Thus, it is difficult to convert from counts in the range 0-127 back to radiances. This point is of considerable significance in the context of change measurement, for any particular count cannot represent a constant spectral radiance from one time to another as the sensor gains and offsets change. If such comparisons cannot be made, then change detection becomes impossible. Furthermore, counts derived from different sensor systems or from different bands of the same image set cannot be compared quantitatively unless they are converted to equivalent spectral radiance values. Radiometric properties of US-processed LANDSAT MSS data are considered in detail by Markham and Barker (1987).

Radiometric calibration of image data is an essential first step in studies of change. Data derived from different sensors, or from the same sensor at different times, are supplied in terms of quantized counts, which must be converted to physical values (wm^{-1} sr^{-1}) if they are to be used in modelling or in studies of change. This is necessary in order to ensure consistency and comparability over time.

Atmospheric correction

Solar radiation incident upon the Earth's surface must pass through the atmosphere. Equally, reflected or emitted radiation passes through and is, in turn, spectrally modified by its interactions with the atmosphere. As the nature of the ground surface cover is inferred from the changes in the spectrum produced by the interaction between incident solar irradiation and surface cover materials, it is necessary to realize that atmospheric interactions also modify the signal. This may not be an important consideration if relative spatial differences between ground surface materials at one point in time are the object of study. However, if images are to be compared over time and inferences made concerning changes in the type, vigour or other property of vegetation, then such changes cannot be reliably confirmed if the effects of the atmosphere are not taken into consideration. Atmospheric constituents (gases and aerosols) scatter and absorb solar radiation. Kaufman (1989) notes the following effects produced by interaction with the atmosphere: variation in the severity of the effect with wavelength, which may affect discrimination between stressed and unstressed vegetation; alteration in the spatial distribution of reflected radiation, affecting the spatial resolution of the system; changes in apparent brightness of a target, affecting measurements of albedo and reflection; and generation of spatial variations in the apparent surface reflection through the effect of subpixel-sized clouds.

Several methods for the atmospheric correction of remotely-sensed images have been reported in the literature; none is universally applicable. As Kaufman (1989) notes:

> The basic philosophy of the atmospheric correction is to obtain information about the atmospheric optical characteristics and to apply this information in a correction scheme. One way to describe this information is by the aerosol optical thickness, phase function, the single-scattering albedo, and the gaseous absorption. For high-resolution imagery, some information about the aerosol vertical profile is also required. The problems in the atmospheric correction are due to the difficulty in determining these characteristics.

In the absence of information about the state of the atmosphere at the time of imaging, some authors, for example, USGS (1979), use as a global estimate of 'haze' or atmospheric path radiance, the minimum value in the image histogram. This value is subtracted from all pixel values in the image. It is acceptable only if the target producing this minimum reflectance is constant from image to image. A similar method, based on the very low surface reflectance of deep clear water in the red wavelengths of the visible spectrum, has been found to be acceptably accurate over oceans (Gordon *et al.*, 1983). A more comprehensive approach to the problem of atmospheric correction is described by Hill and Sturm (1988) which uses the histogram-minimum method to estimate atmospheric path radiance, and obtains estimates of optical depth for each spectral band from aerosol phase functions describing the scattering characteristics of various aerosol types. Corrections for absorption by ozone, carbon dioxide and water vapour are derived from the LOWTRAN model (Kneisys *et al.*, 1983). An approximation to the inherent target reflectance can then be derived for each spectral band as described by Megier *et al.* (1991). Until a reliable and generally-applicable method of correcting images from the solar and thermal wavebands is developed, change detection using remote sensing will remain qualitative rather than quantitative.

Correction for illumination and viewing geometry

Each remote sensing platform has its own specific orbital characteristics; for example, images from LANDSAT and SPOT are collected at 0945 and 1030 respectively. The overpass times of the two NOAA satellites, which carry the AVHRR sensor, are 0230, 0730, 1430 and 1930. For daytime passes of these satellites, solar elevation angles will vary depending on the time of day. The sensors themselves have differing optical characteristics. LANDSAT TM and MSS sensors have a narrow nadir-pointing field of view, whereas AVHRR has a field of view of 57°. SPOT's HRV sensor is pointable, and can be tilted up to $\pm 23°$ from nadir. When these variations are considered in conjunction with characteristics such as the slope and aspect of the land area being surveyed, it is clear that the different sensors are not measuring upwelling radiance under the same conditions. Even if the effects of topographic shadow are excluded from consideration, it is still the case that variations in illumination and viewing angle, together with variations in surface slope and aspect, lead to substantial inter-sensor differences in the levels of recorded radiance, even if the land cover type is constant. Reflection of electromagnetic radiation from plant canopies is direction-dependent; both viewing and illumination directions must be considered, hence the term 'hemispherical bidirectional reflectance factor' (BDRF) which is the distribution, over the hemisphere, of the ratio of radiant exitance to irradiance for a discrete set of viewing and illumination angles. In the case of a specular reflector, angles of incidence and reflection or scattering are equal, whereas for a Lambertian scatterer the distribution of radiant exitance is isotropic. Vegetated surfaces may for convenience, be treated as Lambertian, but in practice this is rarely, if ever, the case. Hence, correction of image data must take account of variations in solar zenith angle (for repetitive viewing from the same platform), view angle variations both within an image and from one image to another, and topographic effects.

Kowalik and Marsh (1982) outline a method for the adjustment of image pixel values for variations in solar zenith angle. They find a linear relationship between pixel value and the cosine of the zenith angle. Royer *et al.* (1985) use data from an airborne system with a 37° scan angle range; they report a significant variation in detected radiance with view angle. These variations are not symmetrical about the nadir and are affected by solar zenith angle. Holben and Justice (1980) consider the effects of ground surface slope angle and aspect variations. The magnitude of the effects was found to depend on the solar zenith angle, the orientation of the slope and its inclination. They conclude that '... the topographic effect on (LANDSAT) data can produce a wide variation in the radiances associated with a given cover type' (Holben and Justice, 1980). Other, more recent studies provide more comprehensive surveys of this topic (Holm *et al.*, 1989; Proy *et al.*, 1989; Moran *et al.*, 1990). It is clear that comparisons between images acquired at different times must be corrected for solar illumination effects, while corrections should be applied to images collected by different sensors to take account of view angle and slope effects if realistic estimates of change over time are to be made.

Georeferencing

In geographical studies it is generally the case that the location of an object, both absolute and relative, is as important as the properties of that object. A knowledge of the location on the Earth's surface of a given pixel is needed if the purpose of the study is to measure temporal changes in the properties inferred from the multispectral pixel values at that location. Equally, knowledge of the position of a given pixel is required if image data are to be integrated with other forms of spatial data. Remotely-sensed images are not maps; their coordinate system (the scan line and row, using matrix conventions) has in general only a weak correspondence with latitude and longitude. Given a knowledge of the orbital geometry of the platform and of the time of imaging it is possible to compute the latitude and longitude of any pixel in the image, to a level of precision dependent on the completeness of knowledge and the spatial resolution of the sensor. Precise orbital positions are not available for the current generation of polar-orbiting satellites, though the orbit of the NOAA satellites is known sufficiently well for the approximate calculation of the latitude and longitude of the rather coarse AVHRR pixels, which range in size from 1.1 km at nadir to 5 km or more at the edge of the image. Even so, accurate georeferencing of AVHRR images requires the use of methods which relate points on the image to points with a known location on the Earth's surface. These latter points are known as ground (or geodetic) control points, and their positions can be found either from maps or from the use of satellite position fixing systems such as GPS. The same technique is used to perform geometrical transformations on medium-resolution images such as those acquired by LANDSAT and SPOT; it is described by Mather (1987).

Conclusions

The move towards inter-disciplinary studies in the Earth and environmental sciences is a recognition of the complementarity of research results in the various component fields of study. A parallel development is also taking place in the technology of spatial data handling, in the form of the geographical information system. The extraction of information from remotely-sensed image data is best accomplished using such systems which offer a variety of tools for spatial analysis, data integration and modelling. However, before remotely-sensed data can be confidently used within a GIS their deficiencies must be first of all identified and then corrected. In particular, the differences in calibration between sensors, and over time for the same sensor, must be corrected if comparability—which is essential in studies of change—is to be achieved. Secondly, variations in the signal that are attributable to the atmosphere must be quantified and removed, for the same reasons. Thirdly, spatial variability in the image that is due to illumination variations and viewing geometry must be subtracted to reveal the true reflectance map and, finally, the data should be geometrically corrected. This last step is required in order to ensure that remotely-sensed data are referenced with respect to the same coordinate system as that used by other spatial data sets within the GIS.

This paper has concentrated on the deficiencies present in remotely-sensed image data that are available in standard form on magnetic tape. Other papers in this volume have revealed the wealth of environmental data that can be extracted from remotely-sensed data, using appropriate techniques. Regional and continental-scale observation and monitoring is now near-operational, while modelling studies which use Earth observation data as the main input representing the dynamic, changing features of the Earth's surface, are becoming a focus of research. The move towards the synergistic use of remotely-sensed data within a GIS can only help to move forward such studies at a more rapid rate.

References

Asrar, G. (Ed.), 1989, *Theory and Applications of Optical Remote Sensing*, New York: John Wiley.

Dickinson, R.E., 1984, Modelling evapotranspiration for three-dimensional global climate models, *Geophysics Monographs*, American Geophysical Union, **29**, 58-72.

Duggin, M.J. and Robinove, C.J., 1991, Assumptions implicit in remote sensing data acquisition and analysis, *International Journal of Remote Sensing*, **11**, 1669-94.

Freden S.C. and Gordon, F. Jr., 1983, LANDSAT Satellites, in Colwell, R.N. (Ed.), *Manual of Remote Sensing, Second Edition, Volume 1: Theory, Instruments and Techniques*, Falls Church, Virginia: American Society of Photogrammetry and Remote Sensing, 517-70.

Gordon, H.R., Clarke, D.K., Brown, J.W., Brown, O.B., Evans, R.H. and Broenkow, W.W., 1983, Phytoplankton pigment concentration in the middle Atlantic Bight: comparison of ship determination and CZCS estimates, *Applied Optics*, **22**, 20-36.

Hill, J. and Sturm, B., 1988, Radiometric normalisation of multi-temporal thematic mapper data for the use of greenness profiles in agricultural landcover classification and vegetation monitoring, in *Proceedings of the EARSeL 8th Symposium: Alpine and Mediterranean Areas*, held in Capri, Italy, 17-20 May 1988, 21-40.

Holben, B.N. and Justice, C.O., 1980, The topographic effect on spectral response on nadir-pointing sensors, *Photogrammetric Engineering and Remote Sensing*, **46**, 1191-200.

Holm, R.G., Jackson, R.D., Yuan, B., Moran, M.S., Slater, P.D. and Biggar, S.F., 1989, Surface reflectance factor retrieval from thematic mapper data, *Remote Sensing of Environment*, **27**, 47-57.

Kaufman, Y., 1989, The atmospheric effect on remote sensing and its correction, in Asrar, G. (Ed.), *Theory and Applications of Optical Remote Sensing*, New York: John Wiley, 336-428.

Kneisys, F.X., Shettle, E.P., Gallery, W.O., Chetwynd, J.H., Abreu, L.W., Selby, J.E.A., Clough, S.A. and Fenn, R.W., 1983, *Atmospheric Transmittance/Radiance: Computer Code LOWTRAN 6*, AFGL-TR-83-0187, Hanscom Air Force Base, Massachusetts: Air Force Geophysics Laboratory.

Kowalik, W.S. and Marsh, S.E., 1982, A relation between LANDSAT digital numbers, surface reflectance, and the cosine of the solar zenith angle, *Remote Sensing of Environment*, **12**, 39–55.

Legg, C.A., 1991, A review of LANDSAT MSS image acquisition in the United Kingdom, 1976-1988, and the implications for operational remote sensing, *International Journal of Remote Sensing*, **12**, 93–106.

Markham B.L. and Barker, J.L., 1987, Radiometric properties of US processed MSS data, *Remote Sensing of Environment*, **22**, 39–71.

Mather, P.M., 1987, *Computer Processing of Remotely-Sensed Images*, Chichester: John Wiley.

Megier, J., Hill, J. and Kohl, H., 1991, Land-use inventory and mapping in a mountainous area: the Ardeche experiment, *International Journal of Remote Sensing*, **12**, 445–62.

Moran, M.S., Jackson, R.D., Hart, G.F., Slater, P.N., Bartell, R.J., Biggar, S.F., Gellman, D.I. and Santer, R.P., 1990, Obtaining surface reflectance factors from atmospheric and view angle corrected SPOT-1 HRV data, *Remote Sensing of Environment*, **32**, 203–14.

Peterson, D.L. and Running, S.W., 1989, Applications in forest science and management, in Asrar, G. (Ed.), *Theory and Applications of Optical Remote Sensing*, New York: John Wiley, 429–73.

Proy, C., Tanré, D. and Deschanps, P.Y., 1989, Evaluation of topographic effects in remotely sensed data, *Remote Sensing of Environment*, **30**, 21–32.

Rhind, D.W. and Green, N.P.A., 1988, Design of a Geographical Information System for a heterogeneous scientific community, *Proceedings of the GIS Conference*, 8–9 June 1988, British Geological Survey, Keyworth, Nottingham. Volume 2: Selected Papers, Swindon: Natural Environment Research Council, 39–57.

Royer, A., Vincent, P. and Bonn, F., 1985, Evaluation and correction of viewing angle effects on satellite measurements of bidirectional reflectance, *Photogrammetric Engineering and Remote Sensing*, **51**, 1899–914.

Running, S.W. and Coughlan, J.C., 1988, A general model of forest ecosystem processes for regional applications. I: Hydrological balance, canopy gas exchange and primary processes, *Ecological Modelling*, **42**, 125–54.

Sellers, P.J. and Dorman, J.L., 1987, Testing the simple biosphere model (SiB) using point micrometeorological and biophysical data, *Journal of Climatology and Applied Meteorology*, **26**, 622–51.

USGS, 1979, *LANDSAT Data Users Handbook*, Sioux Falls, S. Dakota: Eros Data Center.

Wickland, D.E., 1989, Future directions for remote sensing in terrestrial ecological research, in Asrar, G. (Ed.), *Theory and Applications of Optical Remote Sensing*, New York: John Wiley, 691–724.

Wilson, M.F., Henderson-Sellers, A., Dickinson, R.E. and Kennedy, P.J., 1987, Sensitivity of the Biosphere-Atmosphere Transfer Scheme (BATS) to the inclusion of variable soil characteristics, *Journal of Climatology and Applied Meteorology*, **26**, 341–62.

Chapter 18
Automated knowledge-based segmentation of SAR images

S. QUEGAN, R. CAVES and P. HARLEY

Department of Applied and Computational Mathematics,
University of Sheffield,
Sheffield S10 2TN

Introduction

Extraction of information from remotely-sensed data is often carried out without any knowledge of structure in the imagery; pixel-based classification is an obvious example. For many purposes, however, it is necessary to recognize structures, such as agricultural fields or ice floes before the pertinent questions (has it changed? has it moved?) can be formulated. Segmentation is the process of splitting an image into 'homogeneous' regions and hence imposing a geometrical and topological structure on the pixels within it. The human vision system is adept at this operation. Indeed, we cannot prevent ourselves from segmenting the visual input. It appears to be an essential part of image understanding, possibly since it reduces the post-retinal data rate to an extent that is manageable by the successive visual processing (Resnikoff, 1989). Survival requires organization and selective omission of data.

Organization and omission of data are likely to be key issues in the use of the coming generation of space-borne SARs. The very strengths of SAR for terrestrial monitoring (high resolution, reliable data gathering, large-scale cover) create a major problem in data handling and interpretation. Segmentation may provide a means of data compression, distilling the essential information from the SAR imagery, especially over land.

There are several other reasons why segmentation is a desirable objective. It allows image representations which are robust to radiometric and geometric distortions, because such distortions do not affect such topological properties as connectedness and adjacency, and shape is also preserved unless the local distortion is very bad. By emphasizing areas, it removes the need for exact pixel-pixel matching (and hence interpolation) when comparing images. This emphasis on areas is in fact essential for SAR, since speckle causes the information per pixel to be small, and averaging over areas is necessary to extract reliable parameters for extended targets. For many such targets, the only information we can justifiably extract is a single number (the mean of the Rayleigh distribution). For textured regions two numbers suffice (the mean, and the order parameter of the k-distribution), together with a description of the auto-correlation properties of the backscatter. The number of pixels making up the target only affects the standard error of estimates of these parameters. Segmentation ensures that this averaging occurs over proper image units rather than across boundaries. Segmentation is thus an invaluable (and possibly essential) aid to classification and change detection in SAR images.

Segmentation and knowledge

To enable a machine to segment an SAR image, it has to be told or taught certain things. At the most basic level, it needs to know what a segment is. A working definition might describe a segment as 'a statistically homogeneous connected region surrounded by edges, where an edge is a change in a local statistic'. The brain clearly works with a less stringent definition, in particular as regards edges. It exploits visual cues containing large amounts of information (line segments, corner structures) to override local effects. These can be used to continue incomplete edges, to disregard local variations in favour of large-scale organization (reclassifying such local variation as texture) or to allow tolerance to slow changes in local brightness. For example the SAR image shown in Figure 18.1 contains various incomplete edge features, but our visual process manages to fill in the gaps, such that these features are seen as being continuous. Also, the strong edge feature at the lower left allows us to recognize that the textured region below it is a single segment, not many small segments. There is clear physiological evidence (de Valery and de Valery, 1988) that the eye carries out considerable low-pass filtering of the scene, at a range of length scales, and that multiple scales are handled hierarchically. Also, the eye responds only to spatio-temporal change and is sensitive to the local frequency content of the retinal image. At a higher level, the brain appears to impose a 'world model' in which large parts of the scene are classified as homogeneous, as evidenced by the 'pattern continuation' phenomenon (Resnikoff, 1989).

Generalized methods of segmenting SAR images at present make only limited appeal to our understanding of the brain's approach, and have tended to concentrate on single aspects of image structuring rather than the unified hierarchical system used by the brain. We can recognize several types of knowledge used in these methods:

(1) Speckle statistics.
(2) Local change or local homogeneity.
(3) Use of multiple scales.
(4) Weighted use of scene features.
(5) A world model.

How these types of knowledge are embedded in segmentation methods is considered below, following a summary of the essential statistical properties of speckle.

Speckle statistics

A comprehensive treatment of speckle would consider amplitude and multi-look speckle but, for conciseness, only one-look intensity speckle is dealt with here. As is well-known, this is exponentially distributed, with probability density function (PDF)

$$p_I(I) = \frac{1}{\sigma} e^{-\frac{I}{\sigma}} \quad I > 0 \tag{1}$$

Both the mean and standard deviation of I have the value σ, and it is this property which creates many of the problems in handling SAR images. Locally the imagery suffers rapid changes, which have a disastrous effect on

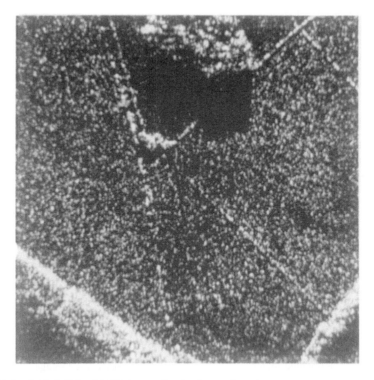

Figure 18.1   RSRE X-band SAR image showing incomplete edges, corner structures, textured regions and other visual cues which are used by the human visual system when segmenting an image.

simple measures of local derivative or local homogeneity (Touzi *et al.*, 1988; Geiss, 1985). As a result, edge detection methods applicable to incoherent images (for example, the Sobel and similar operators) tend to produce many false edges, with the problem becoming worse as brighter segments are considered. Local measures of homogeneity based upon differences from the local mean also perform worse in brighter regions. A number of authors have pointed out the need to use ratios in tests of edge strength or homogeneity (Frost *et al.*, 1982; Touzi *et al.*, 1988; Hendry and Quegan, 1985) if this variation in performance with segment brightness is to be avoided. This is equivalent to tests based on a locally adapted measure of standard deviation.

Segmentation methods

A comprehensive survey of all work on SAR segmentation is beyond the scope of this paper, and the selection below is biased towards work carried out in the UK, and often not readily available in the open literature. A number of different approaches have been taken; here, we comment on the following strategies:

(1)   Segment growing using local similarity rules and local bonding.
(2)   Edge detection followed by bonding.
(3)   Multiple-scale edge detection followed by boundary completion,
(4)   Optimization.
(5)   Image—map matching followed by transfer of segments.

Segment growing

In this approach, the SAR image is filtered and then pixels are 'bonded' to neighbours which they resemble. Two pixels are in the same segment if a chain of bonds connects them (see Figure 18.2a). This approach has been

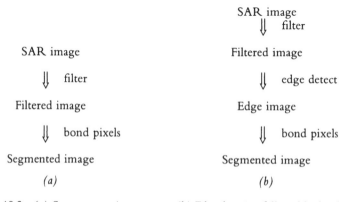

Figure 18.2   (a) Segment:growing.   (b) Edge detection followed by bonding.

studied extensively at Marconi Research Centre, and examples of the results can be found in Quegan *et al.* (1985) and Oddy and Rye (1983). The only knowledge used in this method is of the speckle statistics (this affects the filtering and the criteria for similarity), together with the concept of local homogeneity. Its greatest weakness is its reliance on user-defined parameters at various stages in the process, rather than control by the data itself, and it is consequently hard to quantify its performance.

Edge detection

In this approach, attention is focused on edges, which are the dual concept to segments. Otherwise the procedure is similar to that of segment growing (Figure 18.2b). Some progress has been made to optimize the filtering component of this method and its relation to the edge detector (in fact, the filtering can be absorbed into the edge detector; Hendry and Quegan, 1985). However, more recent work has shown that edge-detection performance may be improved by a multi-scale approach making explicit use of the speckle statistics to assign a probability to any putative edge detection (Touzi *et al.*, 1988; Madsen, 1986). The implications of this for segmentation need to be examined, but this approach is closer to the way the human visual system operates.

Multiple-scale segment growing

Probably the most conceptually satisfying approach to SAR segmentation is that adopted by White (1986), which combines edge detection and a hierarchical scheme for segment growing. An initial edge detection (for known false-alarm probabilities) is used to constrain the search for disc-shaped homogeneous areas. Such areas are placed from large scale down to small scale. Placing of discs at a given scale controls further searches for edges using the inferred knowledge about regions (and hence local variance) in the image. Rules based on similarity control the merging of discs of different scales (Figure 18.3 shows an example of initial steps in the process). This scheme has a number of attractive features. It weights edge detections highly; it operates on a variety of scales; it uses known probabilities of false alarm; and it has been applied successfully (though in a limited study) to change detection (White, 1991).

We can recognize certain weaknesses, however. It tends to break up textured areas into many small segments; it does not use optimal edge detectors; it does not build in the strong visual cues afforded by collinear sets of bright or dark points. More work could be justified in such extensions. Another possible improvement is in the detection of weak edges. The technique is critically dependent on the edge detection step. Interestingly, it does not use speckle-specific edge detectors, relying instead on simple difference operators. In this respect, it may be operating more in the way the

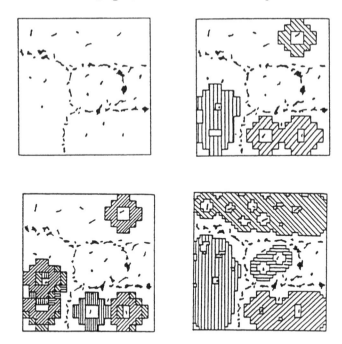

Figure 18.3   Multiple-scale segment growing: an initial edge detection constrains where disks can be placed, starting with large disks and moving down to smaller disks. Similarity rules then control the merging of touching disks.

eye does. It has been observed that morphological filters may act as edge enhancers (see Figure 18.4) which could be applied before the edge detection procedure.

Model-based optimization

The methods described above use very little 'world knowledge', other than imposing conditions on local homogeneity or edge gradients. Work carried out at Liverpool University (Delves *et al.*, 1991) is based on modelling the scene as a collection of uniform areas, i.e. in one dimension the radar reflectivity would be a step function. Segmentation then consists of fitting the data by (two-dimensional) step functions, until a goodness of fit criterion is met. This is computer intensive, but viable.

Segmentation by map matching

An approach to segmentation which makes use of high-level knowledge is possible when the scene to be segmented is of a region where maps exist (or possibly a previously segmented image). This consists of a matching process, followed by a transfer of segments from the map to the image, followed by a test of the reliability of the transferred segmentation. An implicit assumption is that many of the structures in the map will be little changed between the time the map was generated and the time of image formation. Changes will be picked up in the post-transfer processing.

This approach seems to have much to recommend it; it seems unwise to ignore the information in the map, if it can be used; it pinpoints likely problems in less knowledge-based methods, such as the splitting of textured regions into many small segments; it gives a direct link into GIS and into map updating; and it suggests a data–compression scheme based on holding sequences of images as differences from the map segmentation. An indirect gain from this approach, due to its reliance on feature matching, is that it motivates analysis of how features are coded in SAR images.

Feature matching in SAR images has received comparatively little attention (but see Oliver, 1984; Hunting Technical Services and Marconi Research Centre, 1984). When the features to be compared or located are drawn from maps and images, we are faced with the immediate problem of mismatched data types. The map data will be one-bit and if produced by scanning as opposed to vectorization it will contain extraneous symbolic information (it should be possible to devise methods of removing such features) as well as edge information. By contrast, the unpreprocessed SAR

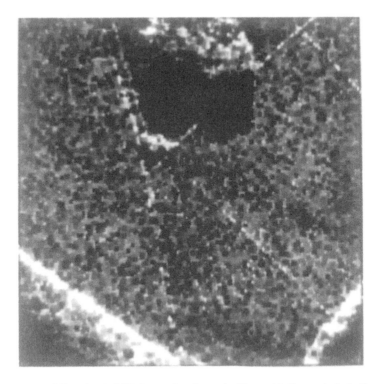

Figure 18.4 RSRE X-band SAR image (as shown in Figure 18.1) morphologically closed using a 5 × 5 window. Closure smooths homogeneous regions whilst preserving edge structures.

data will be grey-scale, and edges and lines may not be well distinguished. The representation of any feature in the SAR image may also be aspect dependent (Hendry et al., 1988).

Any worthwhile algorithm to match map and image data must find an unambiguous match in the right place, which is insensitive to detailed pixel values (within speckle variation). The quality of the match should degrade sharply away from the correct position (weak maxima surrounded by a plateau are likely to be difficult to locate reliably). Matching can be performed on an area basis (Quegan et al., 1988), but here we only describe aspects of matching using boundary information.

When we are searching for a feature defined by boundaries which are bright or dark compared to the surrounding regions, only two types of knowledge seem useful. These are feature geometry, and whether the boundary is dark or bright. We may also be able to use the form of the speckle distribution on the boundary pixels, assuming it consists of a homogeneous population. The first two types of knowledge lead to the simplest matching technique, which is template matching, or correlation of a 1-bit mask with the SAR image, followed by threshold detection. This is equivalent to forming the maximum likelihood (ML) estimate of the local mean in a template shaped like the feature of interest. Viewed in this light, we can use distributional information to form this mean. One way of doing this is to use a Kolmogorov-Smirnov (KS) fit to the data to estimate the mean.

Simulation experiments indicate clearly the superiority of the KS method (Caves et al., 1991), as long as our assumption of a single statistical population in the feature boundary holds good. However, real data is not so amenable. In Figure 18.5 are shown three 100×100 pixel sections of SAR data from an RSRE X-band image, with corresponding sections from digital map data supplied by the OS. From each of the sections of map data has been extracted a 21×21 subsection which is used for the matching procedure. Figure 18.6 shows the results of the attempted match using ML and KS methods. Using ML, only feature 2 is located correctly, while KS finds features 1 and 2 but fails on feature 3.

Examination of the pixel values along the line features in sub-image 3 shows that they cannot be considered as being drawn from a single exponential distribution, so that the extra distributional information which aids the performance of the KS method in simulated data here is false information, and detracts from performance. Neither method performs well enough to act as a reliable feature detector in SAR data.

These results seem to imply that direct methods of map-image matching are unlikely to succeed and it seems necessary to preprocess the imagery. Possibilities which suggest themselves are:

(1) To perform an edge detection on the SAR image first, which has the great advantage of reducing the problem to that of matching compatible (one-bit) data types.

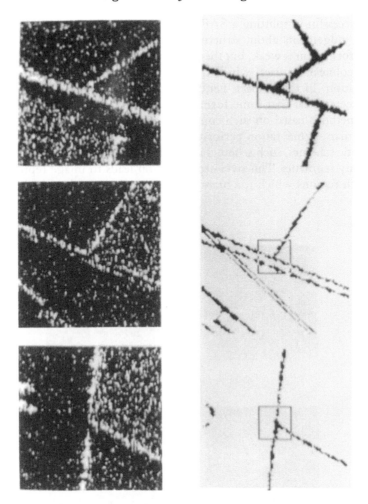

Figure 18.5   Three 100 × 100 pixel sections from a RSRE X-band SAR image and corresponding sections of digital map data; the 21 × 21 pixel areas of map data shown within the square boxes are used as templates for matching the line features in the SAR images.

(2)   To perform a low-level segmentation based on, for example, the White (1986) algorithm, and then to carry out the feature match at this level. Map information may then be fed into the segmentation in order to refine it, and to resolve problems caused by, for example, textured urban or woodland areas.

Discussion

Of the methods of segmentation discussed above, both those based on multiple-scale edge detection and those using an optimization technique seem successful in splitting a SAR image into regions which correspond to human judgements about structure in the image. They tend to fail when edge information is weak, but there are visual cues (such as corners or widely spread colinear features) which the brain can utilize to fill in missing edge information. In this respect, neither method can be regarded as complete, and it seems clear that some form of feature detection may be necessary to guide methods based on such comparatively weak knowledge.

If human segmentation performance is to be our guide on the success of automatic schemes, such schemes also fail in the opposite sense of producing too many segments. This over-segmentation leads to image representations in which regions which the brain would happily regard as single segments

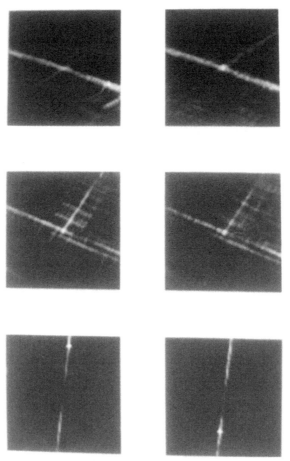

Figure 18.6   Results of matching the features shown in Figure 18.5 using the ML and KS methods (left and right columns respectively); in each case the best match (the maximum) is indicated by a white cross.

become broken into many small segments. Woodland and urban areas where there is marked texture are particularly prone to this affect; lines of trees and hedges are another type of feature which becomes over-segmented. Higher level knowledge about image structure again seems to be needed if emulation of human performance is a real aim.

The feature matching ideas discussed may provide a means to overcome some of these knowledge limitations implicit in 'blind' segmentation. However, the results quoted indicate that feature detection itself may need to use low-level segmentation (or some aspects of it) to generate a match. This interplay and feedback between features and segments needs to be fully exploited if we are to bind the available knowledge sources into segmentation schemes.

Our discussion has concentrated on the technical aspects of segmentation, and its ability to preserve structural information in images at all spatial scales. Any application is likely to significantly constrain the type of information and scales of interest which need to be considered. As a result, very powerful segmentation tools may not be necessary to meet application needs (as an example, see the work on ice-floe tracking described by McConnell *et al.*, 1989).

Indeed, an important aspect of the choice of segmentation algorithm and its implementation for a particular application is dependent on decisions on the data handling structure required to answer the relevant questions. ERS-1 data is likely to provide the first real opportunity to address such issues, especially in the context of forest and ice monitoring. As one application of the above ideas, we intend to investigate the choice of appropriate segmentation techniques for providing high resolution information on changes in forest boundaries over tropical regions during the lifetime of this mission.

Acknowledgements

We would like to thank the Royal Signals and Radar Establishment for the supply of high quality SAR data to support our work and which is used to produce Figures 18.1, 18.3–18.6.

References

Caves, R., Harley, P. and Quegan, S. 1991, Registering SAR images to digital map data using a template matching technique, *Proceedings of the International Geoscience and Remote Sensing Symposium (IGARSS'91)*, **WA11**, New York: IEEE, pp. 1429-32.

Delves, L.M., McQuillan, R.T., Wilkinson, R., Sandys-Renton, J.B.E. and Oliver, C.J., 1991, A two-dimensional segmentation algorithm for SAR images. *Inverse Problems*, **7**, 203-20.

Frost, V.S., Shanmugan, K.S. and Holtzmann, J.C., 1982, Edge detection for synthetic aperture radar and other noisy images, *Proceedings of the International Geoscience and*

Remote Sensing Symposium (IGARSS'82), **FA2**, Noordwijk, The Netherlands: ESA Publications Division, ESTEC, 4.1–4.9.

Geiss, S.C., 1985, *Aspects of SAR Segmentation*, Royal Signals and Radar Establishment, Memo No. 3789.

Hendry, A. and Quegan, S., 1985, Automatic segmentation techniques for SAR images. *Proceedings of the British Pattern Recognition Association Conference*, St. Andrews, pp. 1–10.

Hendry, A., Quegan, S. and Wood, J., 1988, *The visibility of linear features in SAR images*, European Space Agency, SP-284, 1517–20. Noordwijk, The Netherlands.

Hunting Technical Services and Marconi Research Centre, 1984, *Study of land feature extraction from SAR images*, European Space Agency Contract Number 5855/84/GP1(SC).

McConnell, R., Kober, W., Leberl, F., Kwok, R., and Curlander, J., 1989, Automatic tracking of arctic ice flows in multitemporal SAR images, *International Geoscience and Remote Sensing Symposium, (IGARSS'89)*, **2**, 1112–16.

Madsen, S., 1986, *Speckle theory; modelling, analysis and applications related to Synthetic Aperture Radar*. PhD Thesis, Technical University of Denmark.

Oddy, C.J. and Rye, A.J., 1983, Segmentation of SAR images using a local similarity rule, *Pattern Recognition Letters*, **1**, 443–9.

Oliver, C.J., 1984, An analysis of template matching in image registration, *Optica Acta*, **31**, 233–48.

Quegan, S., Veck, N.J., Wright, A., Cruse, D. and Skingley, J., 1985, *The development of automatic methods for handling SAR images over land*, European Space Agency, SP-233, 23–29. Noordwijk, The Netherlands.

Quegan, S., Rye, A.J., Hendry, A., Skingley, J. and Oddy, C.J., 1988, Automatic interpretation strategies for SAR images, *Philosophical Transactions of the Royal Society of London*, **A324**, 409–21.

Resnikoff, H.L., 1989, *The Illusion of Reality*, New York: Springer-Verlag.

Touzi, R., Lopes, A. and Bousquet, P., 1988, A statistical and geometrical edge detector for SAR images, *IEEE Transactions on Geoscience and Remote Sensing*, **GE-26**, 764–73.

de Valery, R. and de Valery, K., 1988, *Spatial Vision*, Oxford Psychology Series No 14, Oxford: Oxford University Press.

White, R., 1986, *Low level segmentation of noisy imagery*, Royal Signals and Radar Establishment, Memo. No. 3900. Malvern.

White, R., 1991, Change detection in SAR imagery, *International Journal of Remote Sensing*, **12**, 339–60.

Chapter 19
Large-scale environmental databases: the example of CORINE

B.K. WYATT

Institute of Terrestrial Ecology,
Monks Wood,
Huntingdon, UK

Introduction

It is usual for human society to organize itself into water–tight compartments or communities. This applies equally to the development of distinct scientific disciplines as to the formation of Departments of State with designated powers and responsibilities. For much of the time, such arrangements are rational and efficient. However, in situations where a response is needed which requires the crossing of disciplinary or institutional boundaries, a compartmentalized social model is probably the worst possible solution; all too often, specialisms create barriers of terminology, or procedure, which appear to be designed with the sole aim of inhibiting meaningful dialogue with groups from other backgrounds.

Environmental questions and issues present obvious examples of situations where an interdisciplinary approach is required. Effective research in the environmental sciences is dependent on cross-fertilization between different individual disciplines. Policies and actions of executive agents of government in respect of the environment must be coordinated to ensure that they take account of existing knowledge and that they are mutually supportive, rather than conflicting in their effects. Whether for scientific research or for environmental protection, a comprehensive and integrated information base is needed, which records states and trends across all environmental media and which documents those human activities which have significant impacts on the environment.

Background

The Stockholm Conference on the Human Environment (Ward and Dubois, 1972), held in the early 1970s and the precursor to the United Nations

233

Environment Programme, gave substance to the call for a more holistic approach to environmental problems and the consequent need for integrated information covering all environmental media. At that time, the pressing global environmental problems were perceived as pollution, desertification and the depletion of non-renewable resources: these priorities were reflected in the early programmes of the major UNEP environmental data initiatives—Global Environmental Monitoring System (GEMS) and Global Resource Information Database (GRID) (Gwynne, 1989).

Simultaneously, the European Community approved and implemented its first Action Programme on the Environment (Commission of the European Communities, 1973). Significantly, the Commission recognized at this early stage the importance of adequate information on the state of the environment and on its sensitivity to change. In 1974, it launched 'Ecological Mapping', a programme intended to provide the Commission's decision-makers with integrated information on the state of the environment in the Community. The Ecological Mapping project was, in many respects, ahead of its time: digital coverage of significant environmental variables was then incomplete and the technology did not exist to allow useful or effective analysis at European scales. Nevertheless, the project established many important principles. Specifically, it proposed the compilation of a database of geographically-referenced information, consistent across all environmental media, and designed to provide a sound basis for formulation of policy at the European level in response to a wide variety of environmental problems (Commission of the European Communities, 1983). Many of these recommendations have now been realized within the CORINE programme which succeeded the Ecological Mapping Project in 1985.

Since this early phase in the move towards integrated environmental information systems, our perceptions of environmental priorities have altered; we now have a better appreciation of the significance of large-scale changes in coupled environmental systems, such as those which determine the global climate. Our understanding of the behaviour of such systems and our ability to predict their future responses is largely determined by the availability of adequate data from experimental studies and from environmental monitoring programmes. Dozier (Chapter 13) has drawn attention to the pivotal importance of data management in the EOS programme (through EOSDIS) and in the International Geosphere-Biosphere Programme (IGBP), through the formation of the IGBP Data and Information System (DIS). In the UK, the Inter-Agency Committee on Global Environmental Change has devoted significant resources to considering the means by which data and information can be made readily available to the global environmental research community.

Despite all this interest and urgent activity, it is fair to remark on the paucity of experience, world-wide, in the establishment of operational multidisciplinary environmental information systems. A Global Database Planning Project was recently established within the framework of IGBP, under the auspices of the International Geographical Union. Its first meeting in 1988 (Mounsey and Tomlinson, 1988), included 20 'Applications Papers', but of these, all but a handful addressed single-discipline themes.

The environmental policy framework in the European Community

The long-term collection of environmental data over large areas is extremely demanding of resources; in the case of European systems, international cooperation is usually needed. For these reasons, large environmental information systems usually exist only in response to clearly defined policy needs. Within the European Community, environmental policy has evolved through a series of Action Programmes, the first dating from as early as 1973 (Commission of the European Communities, 1973). This programme laid down a number of principles on which the Community's environmental policy has subsequently been based. These principles stress:

(1) the importance of preventative measures,
(2) the polluter-pays principle,
(3) the need for international cooperation for control of pollution,
(4) coordination of environmental policies between Member States, without undue interference with national policies,
(5) reconciliation of environmental protection with economic and social development,
(6) avoidance of trans-frontier problems, and
(7) recognition of the needs of Third World countries.

Successive Action Programmes have made explicit reference to the need for effective information to support policy formulation, the assessment of policy options and to monitor the implementation of statutory measures and their effectiveness. The magnitude of this task can be judged by the fact that well over 100 Directives relating to the environment have now been adopted. These Directives, mainly concerned with setting common standards for environmental quality, for emissions and for waste treatment, are often backed up by agreed monitoring procedures and other arrangements for improved dissemination of environmental information.

However, all these measures can be no more than praiseworthy gestures of intent unless the means can be found to ensure that the information needed exists and that it can be made available in suitable forms. In the case of social and economic data, the necessary infrastructure is provided through the Statistical Office of the European Commission, which is responsible for assembling such data from national sources, for integrating them into a Community socio-economic database and for disseminating and publishing Community-wide statistics. Until recently, there was no comparable mechanism for the collection of compatible data on the environment—partly because arrangements for collecting and managing environmental data at the national level are so often piecemeal, incomplete and compartmentalized.

In contrast to this compartmentalization of knowledge and institutions, the environmental media, land, air, the hydrosphere and the biosphere, are interdependent. The environment responds systemically to human activities such as agriculture and industry, while environmental protection measures have impacts both upon environmental quality and on these economic activities. The range of information needed to deal with these complex interactions is potentially enormous. At the European level, the situation is further complicated by the geographical and economic diversity of the region and by the variety of institutional arrangements for collecting and managing environmental information in the different countries.

The more intractable problems associated with the compilation of extensive environmental information are often political, rather than technical, in origin. The establishment of agreed procedures for operating networks for environmental observation and monitoring and agreed standards for recording and exchanging data can present formidable difficulties, especially where these networks cut across national, regional or sectoral boundaries.

However, important elements of an embryonic European environmental information system already exist, in the form of international programmes which manage data on particular topics, as national data compilations which can, in principle, link into international networks and as published documents and maps. It has been the aim of the CORINE programme to exploit and to integrate such sectoral or national data resources, in order to create the Community-wide database needed to support environmental policy and planning at the European level.

The CORINE programme of the European Community

CORINE (CoORdinated INformation on the Environment in the European Community) is a programme of the Environment Directorate of the European Commission, concerned with the development of methods to achieve greater harmonization in the form and content of environmental data and statistics collected within the Member States of the Community. An important spin-off from these initiatives in data harmonization has been the implementation of an experimental information system on the state of the environment and natural resources in the Community. The CORINE programme was initiated in 1985 by a Decision of the Council of Environment Ministers (Commission of the European Communities, 1985) and was funded until the end of 1990.

Objectives

The CORINE programme was intended to address many of the conceptual, technical and political problems associated with the provision of consistent information on environmental states and trends across the Community. Its specific objectives were:

(1) to create a consistent framework for the collection, storage, analysis, presentation and interpretation of environmental data within the Community and to promulgate the adoption of these standards, at the Community level and within Member States,
(2) to collate existing data on the environment in the Community as a whole, within selected thematic areas, and
(3) to coordinate information on the state of the environment and natural resources in the Mediterranean countries. This forms the prototype for a Community-wide system and, at the same time, generates environmental information to support the Community's Integrated Development Programme for the Mediterranean region.

CORINE was conceived as a modular programme (Figure 19.1, after Commission of the European Communities, 1986), which eventually involved more than 200 scientists. The programme comprised the following linked themes:

(1) compilation of a computerized inventory of sources of environmental data in the Community,
(2) design of computer systems for the storage, validation, retrieval and analysis of environmental data, with special emphasis on geographical information systems (GIS),
(3) compilation of basic data on geography, soils and climate to provide a framework for environmental assessment,
(4) compilation of linked data sets covering selected themes, including: land cover,

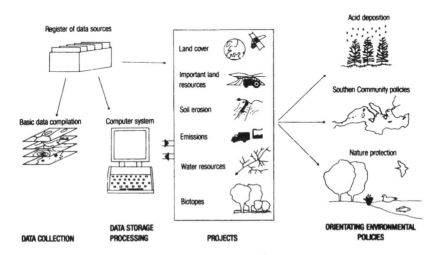

Figure 19.1   The CORINE programme: 1985–1989, after Commission of the European Communities, 1986.

'biotopes' (wildlife and nature), water resources and water quality, emissions to the atmosphere, soil erosion risk and other natural hazards, and

(5)   use of these data sets to support Community environmental policy, particularly in relation to acid deposition, regional development in Mediterranean countries and the protection of wildlife and nature.

Design principles

Two fundamental principles are that the system should exploit existing data resources as far as possible and that, where appropriate, data should be suitable for multiple uses. These design principles were dictated by the expense and difficulty of initiating Europe-wide programmes of environmental data acquisition; in several important respects, as a result of shortcomings in available digital data sets, the implementation has fallen short of these goals.

It is inevitable that data which are collected by national agencies, primarily for national purposes, will exhibit differences which reflect established practice. Data from different countries and regions may be recorded in different units, perhaps using different measurement methods. They may be presented at different spatial and temporal scales and non-quantitative information may use different classifications which may be difficult to reconcile.

Major concerns of the programme were therefore:

(1)   the establishment of common formats and methodologies for data storage and handling,
(2)   the vexed question of quality control,
(3)   the development of models and procedures for data interpretation which are tolerant of the limitations in the data,
(4)   the development of methods for handling spatially-referenced data which are sufficiently flexible to permit the use of material in various forms (e.g. raster data sets, point-line-polygon data), at different scales and geographical projections.

Implementation and project management

In addressing these priority tasks, it was clear that specialist expertise was needed, covering each of the thematic areas included in the CORINE programme. Since the Commission had limited technical resources at its disposal, much of the development of CORINE was placed in the hands of external consultants. The general pattern was for a single consultant group to be given responsibility for one thematic area; each consultant then had the task of designing appropriate procedures and identifying suitable data sources. The overall design framework within which these thematic groups operated, was determined by a Scientific Steering Committee; a separate 'Committee of National Experts' ensured that the views of the governments of the Member States were taken into account. A common pattern was for each thematic group to recruit a team of specialists including, for example, those with knowledge of data availability in each Member State.

Structure of digital representation

The thematic content of the CORINE database is summarized in Table 19.1. Data have been acquired in both vector and in raster form. Some data sets record variables associated with a point location. For example, Biotopes are presently located with reference to the centroids of the areas described, while climatic data are associated with the location of the meteorological stations from which they originate; (the intention is eventually for automatic generation of rainfall and temperature contours). Most of the mapped data are recorded in vector form as points, lines and polygons. A minority of data sets (e.g. terrain, atmospheric emissions) were compiled in raster form, but provision is being made for inter-conversion from raster to vector format, so that it will be possible to select the format which is most appropriate for any given application.

In terms of digital data storage, the database (which corresponds to 70–80 map sheets, covering the 22 themes listed in Table 19.1) occupies in excess of 1 Gb. To improve efficiency of access, it is structured as tiles, each of which corresponds to an area 2° in longitude by 1° in latitude. This tiling is transparent to the user.

Accuracy, precision and scale

Ideally, questions of accuracy and precision should be determined by the uses to which the data are to be put. Unfortunately, given CORINE's dependence upon existing primary data sources, its accuracy is largely pre-determined, and varies both between and within thematic overlays. Of particular relevance in this context is the question of spatial scale. In view of the variety of potential applications for CORINE, an early objective was to design a database that was largely independent of scale. However, there are difficulties in achieving this objective. Generalization of mapped data from large to small scales is fraught with problems. Conversely, attempts to superimpose small-scale and large-scale overlays risks results which are at best meaningless and at worst, may give rise to wholly erroneous conclusions. Yet some compromise is necessary, given the diversity of scales at which environmental data are recorded across the Community. In the short term, two parallel databases have been compiled. The target is a comprehensive Community-wide database, at a nominal scale of 1:1 000 000. At present, this target is not attainable across all themes, so meanwhile, a second system has been compiled at a scale of 1:3 000 000, which includes those features which are at present recorded only at the smaller scale.

In addition to these small-scale synoptic databases, a number of supplementary thematic data sets are being assembled at larger scales, where there is a need for more precise spatial referencing. Two important examples of this are: a digital record of the boundaries of areas of importance for nature conservation; and information on land cover, derived from remote sensing.

Table 19.1 *Overview of the contents of the CORINE Information System.*

Theme	Nature of the information	Volume of information description	Mbytes	Resolution/scale
Biotopes	Location and description of biotopes of major importance for nature conservation in the Community	5600 biotopes described, according to approx. 20 characteristics	20.0	Location of the centre of the site
		Boundaries of 440 biotopes computerized (Portugal, Belgium)	2.0	1/100 000
Designated areas	Location and description of areas classified under various types of protection	13 000 areas described according to approx. 11 characteristics (file being completed)	6.5	Location of the centre of the site
		Computerized record of the limits of the areas designated in compliance with article 4 of the EEC/409/79 directive on conservation of wild birds		1/100 000
Emissions into the air	Tons of pollutants (SO_2, NO_x, VOC) emitted in 1985 per category of source: power stations, industry, transport, nature, oil refineries, combustion	1 value per pollutant, per category of source and per region, plus data for 1400 point sources i.e. ± 200.000 values in total	2.5	Regional (NUTS III) and location of large emission sources
Water resources	Location of gauging station, drainage basin area, mean and minimum discharge, period: 1970–1985, for the southern regions of the EC	Data recorded for 1061 gauging stations, for 12 variables	3.2	Location of gauging station
Coastal erosion	Morpho-sedimentological characteristics (4 categories), presence of constructions, characteristics of coastal evolution: erosion, accretion, stability	17 500 coastal segments described	25.0	Base file: 1/100 000 Generalization: 1/1 000 000

Table 19.1 continued

Theme	Nature of the information	Volume of information description	Mbytes	Resolution/scale
Soil erosion risk	Assessment of the potential and actual soil erosion risk by combining 4 sets of factors: soil, climate, slopes, vegetation	180 000 homogeneous areas (southern regions of the Community)	400.0	1/1 000 000
Important land resources	Assessment of land quality by combining 4 sets of factors: soil, climate, slopes, land improvements	170 000 homogeneous areas (southern regions of the Community)	300.0	1/1 000 000
Natural potential vegetation	Mapping of 140 classes of potential vegetation	2288 homogeneous areas	2.0	1/3 000 000
Land cover	Inventory of biophysical land cover, using 44 class nomenclature	Vectorized database for Portugal, Luxembourg	51.0	1/100 000
Water pattern	Navigability, categories (river, canals, lake, reservoirs)	49 141 digitized river segments	13.8 0.3	1/1 000 000 1/3 000 000
Bathing water quality	Annual values for up to 18 parameters, 113 stations, for 1976–1986, supplied in compliance with EEC/76/160 Directive	2650 values	0·2	Location of station
Soil types	320 soil classes mapped	15.498 homogeneous areas	9.8	1/1 000 000
Climate	Precipitation and temperature (other climatic variables: data incomplete)	Mean monthly values for 4773 stations	7.4	Location of station
Slopes	Mean slope per km² (southern regions of the Community)	1 value per km², i.e. 800.000 values	150.0	1/100 000
Administrative units	EC NUTS regions (Nomenclature of Territorial Units for Statistics) 4 hierarchical levels	470 NUTS digitized	0.7	1/3 000 000

Table 19.1 continued

Theme	Nature of the information	Volume of information description	Mbytes	Resolution/scale
Coasts and countries	Coastline and national boundaries (Community and adjacent territories)	62.734 km	0.3 3.2	1/3 000 000 1/1 000 000
Coasts and countries	Coastline and boundaries (planet)	196 countries	1.5	1/25 000 000
ERDF regions	Eligibility for the Structural Funds	309 regions classified	0.01	Eligible regions
Settlements	Name, location, population of urban centres > 20.000 inhab.	1542 urban centres	0.1	Location of centre
Socio-economic data	Statistical series extracted from the SOEC-REGIO database	Population, transport, agriculture, etc.	40.0	Statistical Units NUTS III
Air traffic	Name, location of airports, type and volume of traffic (1985–87).	254 airports	0.1	Location of airport
Nuclear power stations	Capacity, type of reactor, energy production	97 stations, update 1985	0.03	Location of station

In both cases, the working scale is 1:100 000. Generalization of these data to the 1:1 000 000 base-map is planned, but presents a formidable and as yet unsolved technical challenge.

Data exchange

Consistency is important, not only in the collection and storage of data but also in their transfer and exchange. Given the diversity of its data sources and the number and geographic dispersal of its potential users, this is of particular concern for the CORINE programme. Ultimately, the problem needs to be tackled at a wider international level. In the meanwhile, data transfer specifications have been established, based on the ARC/INFO export format. This is the preferred vehicle for data exchange into and out of the CORINE system, though data can be generated to conform with other industrially-supported formats.

Hardware and software platforms

Several computer systems have been used during the experimental programme. The database was initially implemented using ARC-INFO software, running on a DEC VAX-11/750 at Birkbeck College, University of London. A Siemens SICAD system was subsequently installed in the CORINE offices in Brussels, pending selection of a standard graphics workstation for use throughout the Commission. Other systems were used by individual consultants for data capture. For example, much of the analysis of remotely-sensed imagery required to compile the CORINE land cover map, was carried out using specialist image processing equipment.

The system that is currently used to support the operation of the central CORINE database in Brussels consists of a network of UNIX workstations, running ARC-INFO.

Access conditions

Conditions under which the database may be accessed are still under review. In general, data which are held in CORINE in a processed form (for example, maps of land quality derived from a combination of soil, hydrological and climatic data) are considered to be freely available; the raw data would be accessible only by agreement with the original source. Users in national agencies and organizations which have contributed to data held in CORINE are encouraged to access the system, since it is through use of the data that problems can best be identified and solved. Other users are normally given access to the data by individual agreement. However, it must be recognized that CORINE is still under active development. The database contains known errors and inconsistencies. All information provided from CORINE therefore carries a statement which disclaims responsibility for the consequences of using the data. This disclaimer will remain until the process of quality checking and verification is complete.

Results

The principal result of the CORINE programme has been to successfully demonstrate the feasibility of compiling a consistent geographically-referenced environmental database covering the 2.25 million square kilometres of the European Community. Perhaps of more lasting significance than the database itself is the human and institutional framework on which it has been based—the expert groups and the basic methodologies: on this framework will depend the continuation and the integrity of the programme in future operational contexts.

The CORINE programme has been particularly successful in collaborating with other international groupings in agreeing common procedures and nomenclatures, which in several instances, are gaining acceptance as international standards. Notable examples of this include the principles adopted in collaboration with OECD for CORINAIR and the methods for defining and describing sites of international importance for nature conservation, developed in collaboration with the Council of Europe as part of the CORINE Biotopes Project.

Many of the principles which underpin the CORINE database are being built in to recent Community environmental legislation; for example, the classification used to describe ecological habitats forms the basis of a draft Directive on protection of habitats. In several instances, the CORINE programme has acted as a catalyst in stimulating and accelerating the creation of national environmental information systems; there are many examples where CORINE projects have provided the opportunity to undertake the first coherent national inventories of environmental resources such as biotopes and land cover.

The project provides the basis for a number of important examples of technology transfer. CORINE has been adopted as the basis for initiatives in setting up databases to record biodiversity in several African countries; the World Bank is considering the possibility of applying CORINE methodology to support the assessment of environmental impacts of development projects in Africa; in Central Europe, investment through the PHARE regional programme of the European Community will be linked with an environmental programme which, in large measure, will draw on CORINE for its survey, monitoring and appraisal methods.

Finally, the CORINE programme has provided information and guidelines which have helped in the definition of the structure and aims of the European Environment Agency and has established a sound technical foundation on which to base this innovative Community institution.

Problems and issues

It is hardly surprising that a project of the magnitude of the CORINE programme has encountered problems; these have been scientific, technical and administrative in origin. Important limitations arise from practical constraints on data availability and the difficulty of harmonizing those data which do exist at national or Community level. It is important that users of the CORINE system are aware of such limitations, if the data are not to be misused and misinterpreted. Therefore, great emphasis has been placed on the importance of full and accurate documentation of data. A technical handbook has been published, describing the provenance, completeness and characteristics of each data set in the interim database (Commission of the European Communities, 1989), and a computerized register of data sources is in preparation.

The problems of achieving cartographic consistency between different databases have been legion, and their solution has required the investment of many hours of laborious manual editing. A particularly common, but unforseen difficulty was the absence of basic information about the coordinate reference system, and sometimes the projection used for published maps. As a result, the task of overlaying data sets which at first sight appeared straightforward, was often extremely time-consuming.

The lack of Community-wide data to common standards was a common problem and was particularly acute in the case of basic cartographic data. A digital topographic data set from military sources (Production of Automated Charts—PACE) provides a useful digital record of many important geographical features at an appropriate scale (1:500 000), but covers only the northern half of the Community. There is no corresponding data set for the Mediterranean region. Therefore a less detailed map of coastal outlines and the surface water pattern compiled by Institüt für Angewandte Geodäsie (IFAG) (Brown and Fuller, 1985) was used as a topographic base. (Even the IFAG database did not cover Greece, so that the Greek coastline and rivers had to be separately digitized.) Standardization of data formats has also provided interesting problems and examples of human perversity. In addition to the standards for data exchange mentioned earlier, each of the main thematic activities necessitated the specification of standards for recording the various data. It rapidly became clear that the specification and documentation of such standards is one thing; achieving adherence to those standards is quite another matter! In practice, when the problems of data acquisition were acute, it was usually easier to make *ad hoc* arrangements to read and interpret the data than to attempt to enforce the standards. Consequently, most groups associated with the CORINE programme rapidly became adept at writing 'one-off' programmes to read foreign data sets.

Beside these rather fundamental difficulties, the shortcomings of particular GIS systems paled into insignificance. There have been few instances where it has not been possible to generate products requested by a user because of the limitations of a computer system. There have been many such instances where the quality of the data available have prevented a useful response, or have introduced considerable delay.

The European Environmental Agency

Early in 1989, the President of the European Commission, M. Jacques Delors, in a speech to the European Parliament, proposed the establishment of a European Environmental Agency, which would be administratively separate from the Directorates-General of the Commission and would be responsible for managing a European Environment and Observation Network. The Regulation governing the creation of the Agency was approved by the Council of Environmental Ministers in March 1990, having first received the opinion of the European Parliament (Commission of the European Communities, 1990). It will come into operation as soon as the location of the Agency has been decided by Ministers. The Agency is intended to provide the Community and Member States with objective, reliable and comparable information at Community level to help them take necessary measures for environmental protection, to assess the results of these measures and to ensure that the public is adequately informed. It will provide the necessary scientific and technical support to meet these objectives. The Agency is the logical successor to CORINE. The techniques and databases established under CORINE will provide the nucleus for its initial operations. Indeed, many of the CORINE staff have been recruited to form a 'Task Team' to make technical and administrative preparations for the Agency.

The Agency will be the coordinating body for a European Environment Information and Observation Network. This will comprise:

(1)   National Focal Points, nominated by each Member State;
(2)   elements from national environmental information networks;
(3)   'Topic Centres'—centres of excellence, charged with cooperating with the Agency in specific thematic areas and with defining and coordinating arrangements for information exchange within these themes at the Community level.

The role of the Agency will be primarily to coordinate environmental information activities at the Community level. It will avoid replication of existing functions, it will cooperate with other Community Institutions, such as the Joint Research Centre and the Statistical Office of the European Commission and with relevant international organizations such as the European Space Agency and UN Agencies. Participation of European countries from outside the Community is actively encouraged, as a means of avoiding trans-frontier problems.

The Regulation specifies priority areas for the work of the Agency during its first years, as follows:

(1) air quality and atmospheric emissions,
(2) water quality, pollutants and water resources,
(3) the state of the soil, of flora, fauna and biotopes,
(4) land use and natural resources,
(5) waste management,
(6) noise emissions,
(7) chemical substances hazardous for the environment,
(8) coastal protection.

Within two years of the Agency coming into operation, proposals will be presented for a possible extension of its functions, to include:

(1) monitoring of the implementation of Community legislation (in cooperation with the Commission and existing competent bodies in the Member States),
(2) preparing environmental labels and criteria for award to environmentally-friendly products and technologies,
(3) promoting environmentally-friendly technologies,
(4) establishing criteria for assessing environmental impacts.

Initial discussions have taken place on the details of a possible Work Programme. These decisions now await the appointment of the Director and Management Board, which is expected to take effect shortly after the location of the Agency is known.

Conclusions

The CORINE programme has demonstrated that it is possible to create a multidisciplinary environmental database for Europe, drawing largely on existing data, and that this database can usefully contribute to the preparation and implementation of Community environment policy. CORINE has been particularly successful in drawing up agreed standards and protocols where they previously did not exist.

CORINE is not, and does not pretend to be a tool for global environmental research. However, it provides pointers to the form that such tools might take. Without extensive multidisciplinary information systems of this sort, it will be impossible to respond adequately to the many challenges resulting from global environmental change such as those identified elsewhere in this volume.

Acknowledgements

The author has acted as technical consultant to the European Commission in the design of the CORINE programme and wishes to express his gratitude to the Commission for permission to publish this material and to Dr Dorian Moss and Mrs Cynthia Wiggins for their contributions to the work. The views expressed, however, are the personal opinions of the author and do not necessarily represent EC policy.

References

Brown, N.J. and Fuller, R.M., 1985, A difference in scale? Two digital mapping projects, *The Cartographic Journal*, **22**, 77–82.

Commission of the European Communities, 1973, Action Programme of the European Communities on the Environment, *Official Journal of the European Communities*, **C112**, 20 December, 1.

Commission of the European Communities, 1983, *Communication of the Commission to the Council concerning a methodological approach to an information system on the state of the environment and natural resources in the community (1984–1987)*. **COM(83)**, 528. Brussels.

Commission of the European Communities, 1985, Council Decision of 27 June 1985 on the adoption of the commission work programme concerning an experimental project for gathering, coordinating and ensuring the consistency of information on the state of the environment and natural resources in the Community (85/338/EEC), *Official Journal of the European Communities*, **L 176**, 14–17.

Commission of the European Communities, 1986, *The State of the Environment in the European Community, 1986*, **EUR10633**. Brussels.

Commission of the European Communities, 1989, *CORINE Database Manual, Version 2.2*. **XI/247/90**. Brussels.

Commission of the European Communities, 1990, Council Regulation (EEC) No. 1210/90 of 7 May 1990 on the Establishment of the European Environment Agency and the European Environment Information and Observation Network, *Official Journal of the European Communities*, **L 120/1**.

Gwynne, M.D., 1990, Global monitoring, data management and assessment within GEMS and GRID, *Proceedings of the IUFRO/FAO, International Conference and Workshop Global Natural Resource Monitoring and Assessment: Preparing for the 21st Century*, Bethesda: American Society for Photogrammetry and Remote Sensing. pp. 87–97, Venice, 24–30 September 1989.

Mounsey, H. and Tomlinson, R. (Eds), 1988, Building databases for global science, *Proceedings of the First Meeting of the International Geographical Union Global Database Planning Project*, Tylney Hall, Hampshire, UK, 9–13 May, London: Taylor and Francis.

Ward, B. and Dubois, R., 1972, *Only one Earth. The care and maintenance of a small planet*, an unofficial report commissioned by the Secretary-General of the United Nations Conference on the Human Environment, London: Penguin Books.

Index